# THE MACHINE
## THAT CHANGED THE
# WORLD

BASED ON THE
MASSACHUSETTS
INSTITUTE
OF
TECHNOLOGY
5-MILLION-DOLLAR
5-YEAR
STUDY
ON THE
FUTURE
OF THE
AUTOMOBILE

# THE MACHINE THAT CHANGED THE WORLD

JAMES P. WOMACK
DANIEL T. JONES
DANIEL ROOS

*RAWSON ASSOCIATES*
*New York*

**COLLIER MACMILLAN CANADA**
TORONTO

**MAXWELL MACMILLAN INTERNATIONAL**
NEW YORK   OXFORD   SINGAPORE   SYDNEY

Copyright © 1990 by James P. Womack, Daniel T. Jones, Daniel Roos, and Donna Sammons Carpenter

All rights reserved. No part of this book may be reproduced or transmitted in any form or by any means, electronic or mechanical, including photocopying, recording or by any information storage and retrieval system, without permission in writing from the Publisher.

Rawson Associates
Macmillan Publishing Company
866 Third Avenue, New York, N.Y. 10022
Collier Macmillan Canada, Inc.

Collier Macmillan Canada, Inc.
1200 Eglinton Avenue East
Suite 200
Don Mills, Ontario M3C 3N1

ISBN: 0-89256-350-8
LCC number: 89-063284

Macmillan books are available at special discounts for bulk purchases for sales promotions, premiums, fund-raising, or educational use.
For details, contact:

Special Sales Director
Macmillan Publishing Company
866 Third Avenue
New York, N.Y. 10022

Packaged by Rapid Transcript, a division of March Tenth, Inc.

Designed by Stanley S. Drate/Folio Graphics Co. Inc.

10 9 8 7 6 5 4 3 2 1

Printed in the United States of America

# Contents

ACKNOWLEDGMENTS ........................ *vii*

BEFORE YOU BEGIN THIS BOOK ............. 3

**1** THE INDUSTRY OF INDUSTRIES IN TRANSITION ............................ 11

## THE ORIGINS OF LEAN PRODUCTION ..... 17

**2** THE RISE AND FALL OF MASS PRODUCTION ... 21

**3** THE RISE OF LEAN PRODUCTION ........... 48

## THE ELEMENTS OF LEAN PRODUCTION .... 71

**4** RUNNING THE FACTORY .................. 75

**5** DESIGNING THE CAR ..................... 104

| | | |
|---|---|---:|
| **6** | COORDINATING THE SUPPLY CHAIN | 138 |
| **7** | DEALING WITH CUSTOMERS | 169 |
| **8** | MANAGING THE LEAN ENTERPRISE | 192 |

# DIFFUSING LEAN PRODUCTION 223

| | | |
|---|---|---:|
| **9** | CONFUSION ABOUT DIFFUSION | 227 |
| **10** | COMPLETING THE TRANSITION | 256 |

| | |
|---|---:|
| EPILOGUE | 276 |
| END NOTES | 279 |
| APPENDIXES | |
| **A** INTERNATIONAL MOTOR VEHICLE PROGRAM SPONSORING ORGANIZATIONS | 291 |
| **B** INTERNATIONAL MOTOR VEHICLE PROGRAM RESEARCH AFFILIATE TEAM | 293 |
| **C** IMVP PROGRAM AND FORUM PARTICIPANTS | 295 |
| **D** IMVP PUBLICATIONS LIST | 302 |
| INDEX | 315 |

# Acknowledgments

Writing a book based on the findings of a major research effort and developing a new automobile utilizing many different technologies and manufacturing techniques present many of the same problems. In particular they require a tightly knit project team, clear leadership, and the willingness of many specialists to contribute their personal knowledge and insights to a group effort.

The team putting this book together has been led by Jim Womack, Dan Jones, and Dan Roos, the three senior managers of the IMVP. They have been assisted on a daily basis by John O'Donnell, the IMVP program manager, who has shared his vast knowledge of the auto industry, and by Ann Rowbotham, the IMVP program secretary, who has mastered every detail of a complex project over five years.

The other members of the team have been Donna Carpenter and her associate Abby Solomon, our editorial advisors; Helen Rees, our literary agent; and Eleanor Rawson, our editor at Rawson Associates.

For each of us this has been a sometimes trying but ultimately rewarding experience. Womack, Jones, and Roos had never tried to write for a broad audience, while Carpenter, Rees, and Rawson, with years of experience in commercial publishing, initially found many customs of the academic world curious or even incomprehensible. In the end, we have learned much from each other and we hope this hybrid product—based on a rigorous research program but speaking to a general audience—reflects a successful melding of two distinct cultures.

It would all have been for naught except for the extraordinary generosity of our IMVP research affiliates in sharing their knowledge freely. While three of us are listed as authors because we put the words on the page, this has truly been a group effort by individuals from many backgrounds and countries. We have tried to acknowledge their contributions fully at appropriate points in the text and end notes. The reader should always bear in mind that, like the development of a new car in a "lean" auto company, this has been a truly collective endeavor.

>
> DANIEL ROOS, Director, IMVP
> DANIEL T. JONES, European Director, IMVP
> JAMES P. WOMACK, Research Director, IMVP

# THE
# MACHINE
THAT CHANGED THE
# WORLD

# BEFORE YOU BEGIN THIS BOOK

On a sunny afternoon in the fall of 1984, we stood on the granite front steps of the Massachusetts Institute of Technology and pondered the future. We had just concluded an international conference to announce the publication of our previous book, *The Future of the Automobile*,[1] in which we examined the problems facing the world motor-vehicle industry at that time.

Our findings about the automobile itself were guardedly optimistic. We concluded that technical means were on hand to solve the most pressing environmental and energy problems caused by the use of cars and trucks. There were still question marks about the long term, in particular about the "greenhouse" effect caused in part by carbon dioxide spewing from auto tailpipes, but we thought the automobile itself could adapt. However, we were much gloomier about the auto industry and the world economy.

We concluded that the auto industries of North America and Europe were relying on techniques little changed from Henry Ford's mass-production system and that these techniques were simply not competitive with a new set of ideas pioneered by the Japanese companies, methods for which we did not even have a name. As the Japanese companies gained market share, they were encountering more and more political resistance. At the same time, the Western companies didn't seem to be able to learn from their Japanese competitors. Instead, they were focusing their energies on erecting trade barriers and other competitive impediments, which we thought simply delayed dealing with the real issue. When the next economic downturn came, we feared that

North America and Europe would seal themselves off from the Japanese threat and, in the process, reject the opportunity for the prosperity and more rewarding work that these new techniques offer.

We felt that the most constructive step we could take to prevent this development from occurring would be to undertake a detailed study of the new Japanese techniques, which we subsequently named "lean production," compared with the older Western mass-production techniques, and to do so in partnership with all the world's motor-vehicle manufacturers. But how? As we were pondering this question on that sunny afternoon, one of the senior industry executives attending our conference approached us . . . with precisely this idea.

"Why not also include governments worried about revitalizing their motor-vehicle industries," he asked, "and raise enough funds to really do the job properly?" Thus was born the International Motor Vehicle Program (IMVP) at Massachusetts Institute of Technology and, ultimately, this book.

## THE INTERNATIONAL MOTOR VEHICLE PROGRAM

At the beginning of 1985, a fortuitous event at MIT provided the ideal institutional setting for the IMVP. A new Center for Technology, Policy and Industrial Development was formed with Daniel Roos as its first director. The Center had a bold charter: to go beyond conventional research to explore creative mechanisms for industry-government-university interaction on an international basis in order to understand the fundamental forces of industrial change and improve the policy-making process in dealing with change. The IMVP was an ideal program for the new Center to demonstrate a creative role for a university in working cooperatively with governments and industry.

As we moved ahead with planning the IMVP in the new Center, we realized that our success would depend critically on six elements: thoroughness, expertise, a global outlook, independence, industry access, and continuous feedback.

First, we had to examine the *entire* set of tasks necessary to manufacture a car or truck: market assessment, product design, detailed engineering, coordination of the supply chain, operation of individual factories, and sales and service of the finished product. We knew that many efforts to understand this industry had failed because they never looked further than the factory, an important element in the system to be sure, but only a small part of the total.

We realized that to do a thorough job we would need many types of expertise of a sort not normally found in a university setting. We would need researchers knowledgeable about each aspect of the system who were committed to rigorous research methods, but who were also comfortable with the inherent messiness of the industrial world, where nothing is ever as neat as in academic models. Our solution was to find researchers now in academia who had come from the world of industry and who were willing to go back into design shops, supply companies, and factories for weeks or months to gather the detailed information we needed for sound conclusions.

For example, Richard Lamming and Toshihiro Nishiguchi, our specialists in supply systems, were pursuing Ph.D.'s in England at the University of Sussex and Oxford University, respectively, during their tenure with the IMVP. However, their interest in supply came from their previous work experience in Western and Japanese companies. Richard had been a parts buyer for Jaguar in England, while Toshihiro had worked for Pioneer Electric in Japan. During their four years of work for the IMVP they visited hundreds of component supply companies and plants in North America, Western Europe, and Japan. In addition, they examined supply systems in the leading developing countries, including Korea, Taiwan, and Mexico.

Similarly, Andrew Graves, our technology specialist, was pursuing a Ph.D. at the University of Sussex after many years in a career as a builder of Formula 1 racing cars. Andy spent months traveling to the major design and engineering centers of the motor-vehicle world. On each visit he was testing ideas about the best means for companies to introduce new technologies, ideas formed initially in the world of auto racing, where continuous technical leadership is the key to success.

One of our factory specialists, John Krafcik, was the first American engineer hired at the Toyota–General Motors joint venture, NUMMI. His training at NUMMI included lengthy periods in Japan at Toyota factories in Toyota City, where he learned the fundamentals of lean production at the source. John completed an MBA degree at MIT's Sloan School of Management while traveling the world surveying ninety auto assembly plants in fifteen countries, in what we believe is the most comprehensive industrial survey ever undertaken in any industry.

Two additional MBA students at MIT, Antony Sheriff and Kentaro Nobeoka, provided insight for our product-development studies, through case studies of the product-development process based on their previous work as product planners at Chrysler and Mazda, respectively.

A mere listing of these names shows an additional feature of our work that we felt was essential—to develop a completely international team of researchers, with the language and cultural skills to understand production methods in different countries and an eagerness to explain their findings to colleagues from very different backgrounds. These researchers (who are listed in Appendix B) were not primarily stationed at MIT and were not primarily American. Rather, we developed a worldwide team with no geographic center and no one nationality in the majority.

To be taken seriously both inside and outside the motor-vehicle industry we needed to be independent. Therefore, we determined to raise the $5 million we needed through contributions from many car companies, components suppliers, and governments. (The thirty-six organizations ultimately contributing to the IMVP are listed in Appendix A.) We limited contributions from individual companies and governments to 5 percent of the $5-million total and placed all the funds in a single account, so that no sponsor could influence the direction of our work by earmarking its contribution for a special purpose. We were also careful to raise funds in equal amounts in North America, Western Europe, and Japan, so that we would not be subject to national or regional pressures in our conclusions.

For our researchers to succeed, they would need extensive access to motor-vehicle companies across the world, from the factory floor to the executive suite. We therefore made it clear to potential sponsors that their most valuable contribution would not be money but rather the time given by their employees to answer our questions. In every case these companies have been even more open than we had hoped. We have been truly amazed by the spirit of professionalism in this industry, which has moved managers in the worst facilities and weakest companies to share their problems frankly, and managers in the best plants and strongest companies to explain their secrets candidly.

Finally, to succeed in our work we were determined to devise a set of feedback mechanisms where we could explain our findings to industry, governments, and unions and gain their reactions for our mutual benefit. We did this in three ways.

First, we held an annual meeting for the liaison person from each sponsor. At these meetings we went over the previous year's research in detail, asking for criticism and for suggestions about the next steps for our research.

Second, we held an annual policy forum at a different location around the world—Niagara-on-the-Lake in Canada, Como in Italy, Acapulco in Mexico—to present our findings to senior executives and government officials from the sponsoring companies and governments, plus interested observers from labor unions and the financial community. These private meetings provided an opportunity for senior leaders of this industry to discuss the real problems of moving the world from mass to lean production, outside the glare of publicity and the need for public posturing. (Those attending the IMVP policy forums are listed in Appendix C.)

Finally, we've conducted several hundred private briefings for companies, governments, and unions. For example, our factory practice team conducted a seminar at each of the ninety assembly plants we visited as part of the IMVP World Assembly Plant Survey. In these seminars, we reviewed worldwide performance, assessed the performance of the plant we were visiting, and explained the reasons that plant might lag in world-class performance. We also conducted briefings for corporate management boards, union executive committees, government ministries, and leaders in the investment community, in each case explaining the differences between mass production and lean with ideas on how to convert to lean production.

## THIS BOOK

We have now spent five years exploring the differences between mass production and lean production in one enormous industry. We have been both insiders with access to vast amounts of proprietary information and daily contact with industry leaders, and outsiders with a broad perspective, often very critical, on existing practices. In this process we've become convinced that the principles of lean production can be applied equally in every industry across the globe and that the conversion to lean production will have a profound effect on human society—it will truly change the world.

We therefore decided not to write an academic report on our work, a dry summary of findings by a committee seeking a consensus. Instead, in the pages that follow, the three of us, as leaders of the Program, tell the story of how human society went about making things during the rise, and now the decline of the age of mass production, and how some companies in some countries have pioneered a new way of making things in the dawning age of lean production. In the last part of our book, we provide a vision of how the whole world can enter this new age.

Our story draws on the 116 research monographs prepared by IMVP Research Affiliates, as listed in Appendix D, but necessarily provides only a small fraction of the evidence behind our analysis. Readers with further interest in specific topics should consult Appendix D and write for copies to the IMVP, Center for Technology, Policy and Industrial Development, E40-219, MIT, Cambridge, MA 02139 U.S.A.

Readers should realize that with such a rich diversity of global intellectual resources and viewpoints, IMVP researchers have not agreed on every point. This volume presents the personal view of the three Program leaders and should not be taken as an official statement agreed to by all participants. Certainly, they should not be blamed for any errors or omissions.

Our story is not just for an industry audience but for everyone—government officials, labor leaders, industry executives, and general readers—in every country with an interest in how society goes about making things. In the process, we necessarily make some unflattering comparisons of companies and countries. We ask the reader to take these in the proper spirit. We have no wish to embarrass, or for that matter to compliment, but rather to illustrate the transition from mass to lean production with concrete examples that readers can understand.

We also ask the reader to understand that our sponsors have been extraordinarily supportive of our work. They have sent senior executives to our annual meetings and several have given us a critique of a draft of this volume—in some cases voicing disagreements. However, they have neither exercised veto power over our findings nor endorsed our conclusions. The views in the pages ahead are strictly our own. For our sponsors' willingness to let us think big thoughts without interference at a time of profound transition, we are deeply grateful.

## A FINAL CHALLENGE FOR THE READER

In presenting our work to a broad audience we have one great fear: that readers will praise it or condemn it as yet another "Japan" book, concerned with how a sub-set of the population within a relatively small country produces manufactured goods in a unique way. Our intention is emphatically different. We believe that the fundamental ideas of lean production are universal—applicable anywhere by anyone—and that many non-Japanese companies have already learned this.

Thus we devote our attention in the pages ahead to a careful explanation of the logic and techniques of lean production. We pay little attention to the special features of Japanese society—the high savings rate, near universal literacy, a homogeneous population, the often alleged inclination to subordinate personal desires to group needs, and the willingness, even the desire, to work long hours—which some observers credit for Japanese success, but which we believe are of secondary importance.

Similarly, we pay little attention to other features of Japanese society—the limited role for women and minorities in the economy, the tight relation between government and industry, the barriers to foreign penetration of the domestic market, and the pervasive distinction between foreign and Japanese—which other countries adopting lean production would neither want nor need to copy. This is not a book about what is wrong with Japan or with the rest of the world but about what is right with lean production.

Nevertheless, the level of tension about trade and investment between Japan and the rest of the world is now so great that most readers, in Japan as well as the West, will need to make a special effort to extract the universal principles of lean production from their initial, Japanese application.

Early in this century, most Europeans were unable to differentiate the universal ideas and advantages of mass production from their unique American origins. As a result, ideas of great benefit were rejected for a generation. The great challenge of the current moment is to avoid making such an error twice.

# THE INDUSTRY OF INDUSTRIES IN TRANSITION

## 1

Forty years ago Peter Drucker dubbed it "the industry of industries."[1] Today, automobile manufacturing is still the world's largest manufacturing activity, with nearly 50 million new vehicles produced each year.

Most of us own one, many of us own several, and, although we may be unaware of it, these cars and trucks are an important part of our everyday lives.

Yet the auto industry is even more important to us than it appears. Twice in this century it has changed our most fundamental ideas of how we make things. And how we make things dictates not only how we work but what we buy, how we think, and the way we live.

After World War I, Henry Ford and General Motors' Alfred Sloan moved world manufacture from centuries of craft production—led by European firms—into the age of mass production. Largely as a result, the United States soon dominated the global economy.

After World War II, Eiji Toyoda and Taiichi Ohno at the Toyota Motor Company in Japan pioneered the concept of lean production. The rise of Japan to its current economic preeminence quickly followed, as other Japanese companies and industries copied this remarkable system.

Manufacturers around the world are now trying to embrace lean production, but they're finding the going rough. The companies that first mastered this system were all headquartered in one country—Japan. As lean production has spread to North America and Western Europe under their aegis, trade wars and growing resistance to foreign investment have followed.

Today, we hear constantly that the world faces a massive overcapacity crisis—estimated by some industry executives at more than 8 million units in excess of current world sales of about 50 million units.[2] This is, in fact, a misnomer. The world has an acute shortage of competitive lean-production capacity and a vast glut of uncompetitive mass-production capacity. The crisis is caused by the former threatening the latter.

Many Western companies now understand lean production, and at least one is well along the path to introducing it. However, superimposing lean-production methods on existing mass-production systems causes great pain and dislocation. In the absence of a crisis threatening the very survival of the company, only limited progress seems to be possible.

General Motors is the most striking example. This gigantic company is still the world's largest industrial concern and was without doubt the best at mass production, a system it helped to create. Now, in the age of lean production, it finds itself with too many managers, too many workers, and too many plants. Yet GM has not yet faced a life-or-death crisis, as the Ford Motor Company did in the early 1980s, and thus it has not been able to change.[3]

This book is an effort to ease the necessary transition from mass production to lean. By focusing on the global auto industry, we explain in simple, concrete terms what lean production is, where it came from, how it really works, and how it can spread to all corners of the globe for everyone's mutual benefit.

But why should we care if world manufacturers jettison decades of mass production to embrace lean production? Because the adoption of lean production, as it inevitably spreads beyond the auto industry, will change everything in almost every industry—choices for consumers, the nature of work, the fortune of companies, and, ultimately, the fate of nations.

What is lean production? Perhaps the best way to describe this innovative production system is to contrast it with craft production and mass production, the two other methods humans have devised to make things.

The craft producer uses highly skilled workers and simple but

# THE INDUSTRY OF INDUSTRIES IN TRANSITION

flexible tools to make exactly what the consumer asks for—one item at a time. Custom furniture, works of decorative art, and a few exotic sports cars provide current-day examples. We all love the idea of craft production, but the problem with it is obvious: Goods produced by the craft method—as automobiles once were exclusively—cost too much for most of us to afford. So mass production was developed at the beginning of the twentieth century as an alternative.

The mass-producer uses narrowly skilled professionals to design products made by unskilled or semiskilled workers tending expensive, single-purpose machines. These churn out standardized products in very high volume. Because the machinery costs so much and is so intolerant of disruption, the mass-producer adds many buffers—extra supplies, extra workers, and extra space—to assure smooth production. Because changing over to a new product costs even more, the mass-producer keeps standard designs in production for as long as possible. The result: The consumer gets lower costs but at the expense of variety and by means of work methods that most employees find boring and dispiriting.

The lean producer, by contrast, combines the advantages of craft and mass production, while avoiding the high cost of the former and the rigidity of the latter. Toward this end, lean producers employ teams of multiskilled workers at all levels of the organization and use highly flexible, increasingly automated machines to produce volumes of products in enormous variety.

Lean production (a term coined by IMVP researcher John Krafcik) is "lean" because it uses less of everything compared with mass production—half the human effort in the factory, half the manufacturing space, half the investment in tools, half the engineering hours to develop a new product in half the time. Also, it requires keeping far less than half the needed inventory on site, results in many fewer defects, and produces a greater and ever growing variety of products.

Perhaps the most striking difference between mass production and lean production lies in their ultimate objectives. Mass-producers set a limited goal for themselves—"good enough," which translates into an acceptable number of defects, a maximum acceptable level of inventories, a narrow range of standardized products. To do better, they argue, would cost too much or exceed inherent human capabilities.

Lean producers, on the other hand, set their sights explicitly on perfection: continually declining costs, zero defects, zero inventories, and endless product variety. Of course, no lean producer has ever reached this promised land—and perhaps none ever will, but the endless quest for perfection continues to generate surprising twists.

For one, lean production changes how people work but not always in the way we think. Most people—including so-called blue-collar workers—will find their jobs more challenging as lean production spreads. And they will certainly become more productive. At the same time, they may find their work more stressful, because a key objective of lean production is to push responsibility far down the organizational ladder. Responsibility means freedom to control one's work—a big plus—but it also raises anxiety about making costly mistakes.

Similarly, lean production changes the meaning of professional careers. In the West, we are accustomed to think of careers as a continual progression to ever higher levels of technical know-how and proficiency in an ever narrower area of specialization as well as responsibility for ever larger numbers of subordinates—director of accounting, chief production engineer, and so on.

Lean production calls for learning far more professional skills and applying these creatively in a team setting rather than in a rigid hierarchy. The paradox is that the better you are at teamwork, the less you may know about a specific, narrow specialty that you can take with you to another company or to start a new business. What's more, many employees may find the lack of a steep career ladder with ever more elaborate titles and job descriptions both disappointing and disconcerting.

If employees are to prosper in this environment, companies must offer them a continuing variety of challenges. That way, they will feel they are honing their skills and are valued for the many kinds of expertise they have attained. Without these continual challenges, workers may feel they have reached a dead end at an early point in their career. The result: They hold back their know-how and commitment, and the main advantage of lean production disappears.

This sketch of lean production and its effects is highly simplified, of course. Where did this new idea come from and precisely how does it work in practice? Why will it result in such profound political and economic changes throughout the world? In this book we provide the answers.

In "The Origins of Lean Production," we trace the evolution

of lean production. We then look in "The Elements of Lean Production" at how lean production works in factory operations, product development, supply-system coordination, customer relations and as a total lean enterprise.

Finally, in "Diffusing Lean Production," we examine how lean production is spreading across the world and to other industries and, in the process, is revolutionizing how we live and work. As we'll also see, however, lean production isn't spreading everywhere at a uniform rate. So we'll look at the barriers that are preventing companies and countries from becoming lean. And we'll suggest creative ways leanness can be achieved.

# THE ORIGINS OF LEAN PRODUCTION

# THE ORIGINS OF
# LEAN PRODUCTION

No new idea springs full-blown from a void. Rather, new ideas emerge from a set of conditions in which old ideas no longer seem to work. This was certainly true of lean production, which arose in one country at a specific time because conventional ideas for the industrial development of the country seemed unworkable. Therefore, to understand lean production and its origins fully, it is important to go much farther back in time, in fact, back to the origins of the motor industry at the end of the nineteenth century.

In Chapter 2, we look at the craft origins of the industry in the 1880s and the transition to mass production around 1915, when craft production encountered problems it could not surmount. We take pains to describe the mature system of mass production as it came to exist by the 1920s, including its strengths and weaknesses, because the system's weaknesses eventually became the source of inspiration for the next advance in industrial thinking.

In Chapter 3, we are then ready to examine the genesis of lean production in the 1950s and how it took root. We also summarize the key features of the fully developed lean production system as it came to exist in Japan by the 1960s, at a point long before the rest of the world took note.

# THE RISE AND FALL OF MASS PRODUCTION

## 2

In 1894, the Honorable Evelyn Henry Ellis, a wealthy member of the English Parliament, set out to buy a car.[1] He didn't go to a car dealer—there weren't any. Nor did he contact an English automobile manufacturer—there weren't any of those either.

Instead, he visited the noted Paris machine-tool company of Panhard et Levassor and commissioned an automobile. Today, P&L, as it was known, is remembered only by classic-car collectors and auto history buffs, but, in 1894, it was the world's leading car company.[2]

It got its start—and a jump on other potential competitors—when in 1887 Emile Levassor, the "L" of P&L, met Gottlieb Daimler, the founder of the company that today builds the Mercedes-Benz. Levassor negotiated a license to manufacture Daimler's new "high-speed" gasoline engine.

By the early 1890s, P&L was building several hundred automobiles a year. The cars were designed according to the *Système Panhard*—meaning the engine was in the front, with passengers seated in rows behind, and the motor drove the rear wheels.

When Ellis arrived at P&L, which was still primarily a manufacturer of metal-cutting saws rather than automobiles, he found in place the classic craft-production system. P&L's work force was overwhelmingly composed of skilled craftspeople who carefully hand-built cars in small numbers.

These workers thoroughly understood mechanical design principles and the materials with which they worked. What's more, many were their own bosses, often serving as independent contractors within the P&L plant or, more frequently, as independent machine-shop owners with whom the company contracted for specific parts or components.

The two company founders, Panhard and Levassor, and their immediate associates were responsible for talking to customers to determine the vehicle's exact specifications, ordering the necessary parts, and assembling the final product. Much of the work, though, including design and engineering, took place in individual craft shops scattered throughout Paris.

One of our most basic assumptions in the age of mass production—that cost per unit falls dramatically as production volume increases—was simply not true for craft-based P&L. If the company had tried to make 200,000 identical cars each year, the cost per car probably wouldn't have dipped much below the cost per car of making ten.

What's more, P&L could never have made two—much less 200,000—identical cars, even if these were built to the same blueprints. The reasons? P&L contractors didn't use a standard gauging system, and the machine tools of the 1890s couldn't cut hardened steel.

Instead, different contractors, using slightly different gauges, made the parts. They then ran the parts through an oven to harden their surfaces enough to withstand heavy use. However, the parts frequently warped in the oven and needed further machining to regain their original shape.

When these parts eventually arrived at P&L's final assembly hall, their specifications could best be described as approximate. The job of the skilled fitters in the hall was to take the first two parts and file them down until they fit together perfectly.

Then they filed the third part until it fit the first two, and so on until the whole vehicle—with its hundreds of parts—was complete.

This sequential fitting produced what we today call "dimensional creep." So, by the time the fitters reached the last part, the total vehicle could differ significantly in dimensions from the car on the next stand that was being built to the same blueprints.

Because P&L couldn't mass-produce identical cars, it didn't

try. Instead, it concentrated on tailoring each product to the precise desires of individual buyers.

It also emphasized its cars' performance and their hand-fitted craftsmanship in which the gaps between individual parts were nearly invisible.

To the consumers Panhard was trying to woo, this pitch made perfect sense. These wealthy customers employed chauffeurs and mechanics on their personal staffs. Cost, driving ease, and simple maintenance weren't their primary concerns. Speed and customization were.

Evelyn Ellis was no doubt typical of P&L's clients. He didn't want just any car; he wanted a car built to suit his precise needs and tastes. He was willing to accept P&L's basic chassis and engine, he told the firm's owners, but he wanted a special body constructed by a Paris coachbuilder.

He also made a request to Levassor that would strike today's auto manufacturer as preposterous: He asked that the transmission, brake, and engine controls be transferred from the right to the left side of the car. (His reason wasn't that the English drove on the left—in that case, moving the controls to the left side of the vehicle was precisely the wrong thing to do. Besides, the steering tiller remained in the middle. Rather, he presumably thought the controls were more comfortable to use in that position.)

For P&L, Ellis's request probably seemed simple and reasonable. Since each part was made one at a time, it was a simple matter to bend control rods to the left rather than the right and to reverse the linkages. For today's mass-producer, this modification would require years—and millions or hundreds of millions of dollars—to engineer. (American companies still offer no right-side-drive option on cars they sell in drive-on-the-left Japan, since they believe the cost of engineering the option would be prohibitive.)

Once his automobile was finished, Ellis, accompanied by a mechanic engaged for the purpose, tested it extensively on the Paris streets. For, unlike today's cars, the vehicle he had just bought was in every sense a prototype. When he was satisfied that his new car operated properly—quite likely after many trips back to the P&L factory for adjustment—Ellis set off for England.

His arrival in June 1895 made history. Ellis became the first person to drive an automobile in England. He traversed the fifty-six miles from Southampton to his country home in a mere 5 hours and 32 minutes—exclusive of stops—for an average speed of 9.84 miles per hour. This speed was, in fact, flagrantly illegal, since the limit in England for non-horse-drawn vehicles was a sedate 4 miles per hour. But Ellis didn't intend to remain a lawbreaker.

By 1896, he had taken the Parliamentary lead in repealing the "flag law" that limited automotive speeds, and had organized an Emancipation Run from London to Brighton, a trip on which some cars even exceeded the new legal speed limit of 12 miles per hour. Around this time, a number of English firms began to build cars, signaling that the automotive age was spreading from its origins in France to England in its march across the world.

Evelyn Ellis and P&L are worth remembering, despite the subsequent failure of the Panhard firm and the crudeness of Ellis's 1894 auto (which found a home in the Science Museum in London, where you can see it today). Together, they perfectly summarize the age of craft production in the motor industry.

In sum, craft production had the following characteristics:

- A work force that was highly skilled in design, machine operations, and fitting. Most workers progressed through an apprenticeship to a full set of craft skills. Many could hope to run their own machine shops, becoming self-employed contractors to assembler firms.
- Organizations that were extremely decentralized, although concentrated within a single city. Most parts and much of the vehicle's design came from small machine shops. The system was coordinated by an owner/entrepreneur in direct contact with everyone involved—customers, employers, and suppliers.
- The use of general-purpose machine tools to perform drilling, grinding, and other operations on metal and wood.
- A very low production volume—1,000 or fewer automobiles a year, only a few of which (fifty or fewer) were built to the same design. And even among those fifty, no two were exactly alike since craft techniques inherently produced variations.

No company, of course, could exercise a monopoly over these resources and characteristics, and Panhard et Levassor was soon competing with scores of other companies, all producing vehicles in a similar manner. By 1905, less than twenty years after P&L produced the first commercially successful automobile, hundreds of companies in Western Europe and North America were turning out autos in small volumes using craft techniques.

The auto industry progressed to mass production after World War I, and P&L eventually foundered trying to make the conversion. Yet, a number of craft-production firms have survived up to the present. They continue to focus on tiny niches around the upper, luxury end of the market, populated with buyers wanting a unique image and the opportunity to deal directly with the factory in ordering their vehicles.

Aston Martin, for example, has produced fewer than 10,000 cars at its English workshop over the past sixty-five years and currently turns out only one automobile each working day. It survives by remaining small and exclusive, making a virtue of the high prices its craft-production techniques require. In its body shop, for example, skilled panel beaters make the aluminum body panels by pounding sheets of aluminum with wooden mallets.

In the 1980s, as the pace of technological advances in the auto industry has quickened, Aston Martin and similar firms have had to ally themselves with the automotive giants (Ford, in Aston Martin's case[3]) in order to gain specialized expertise in areas ranging from emission controls to crash safety. The cost of their developing this expertise independently would have been simply prohibitive.

In the 1990s, yet another threat will emerge for these craft firms as companies mastering lean production—led by the Japanese—begin to pursue their market niches, which were too small and specialized for the mass-producers, such as Ford and GM, ever to have successfully attacked. For example, Honda has just introduced its aluminum-bodied NS-X sports car, which is a direct attack on Ferrari's niche in ultra-high-performance sports cars. If these lean-production firms can cut design and manufacturing costs and improve on the product quality offered by the craft firms—and they probably can—the traditional craft producers will either have to adopt lean-production methods themselves or perish as a species after more than a century.

Nostalgists see Panhard and its competitors as the golden age of auto production: Craftsmanship counted and companies gave their full attention to individual consumers. Moreover, proud craft workers honed their skills and many became independent shop owners.

That's all true, but the drawbacks of craft production are equally obvious in hindsight. Production costs were high and didn't drop with volume, which meant that only the rich could afford cars. In addition, because each car produced was, in effect, a prototype, consistency and reliability were elusive. (This, by the way, is the same problem that plagues satellites and the U.S. space shuttle, today's most prominent craft products.)

Car owners like Evelyn Ellis, or their chauffeurs and mechanics, had to provide their own on-the-road testing. In other words, the system failed to provide product quality—in the form of reliability and durability rather than lots of leather or walnut—because of the lack of systematic testing.

Also fatal to the age, however, was the inability of the small independent shops, where most of the production work took place, to develop new technologies. Individual craftsmen simply did not have the resources to pursue fundamental innovations; real technological advance would have required systematic research rather than just tinkering. Add these limitations together and it is clear, in retrospect, that the industry was reaching a plateau when Henry Ford came along. That is, as the general design of cars and trucks began to converge on the now familiar four-wheel, front-engine, internal-combustion vehicle we know today, the industry reached a premature maturity, fertile ground for a new production idea.

At this point, Henry Ford found a way to overcome the problems inherent in craft production. Ford's new techniques would reduce costs dramatically while increasing product quality. Ford called his innovative system *mass production*.[4]

## MASS PRODUCTION

Ford's 1908 Model T was his twentieth design over a five-year period that began with the production of the original Model A in 1903. With his Model T, Ford finally achieved two objectives. He had a car that was designed for manufacture, as we would say today, and that was, also in today's terms, user-friendly. Almost anyone could drive and repair the car without a chauffeur or mechanic. These two achievements laid the groundwork for the revolutionary change in direction for the entire motor-vehicle industry.[5]

The key to mass production wasn't—as many people then and

now believe—the moving, or continuous, assembly line. *Rather, it was the complete and consistent interchangeability of parts and the simplicity of attaching them to each other.* These were the manufacturing innovations that made the assembly line possible.

To achieve interchangeability, Ford insisted that the same gauging system be used for every part all the way through the entire manufacturing process. His insistence on working-to-gauge throughout was driven by his realization of the payoff he would get in the form of savings on assembly costs. Remarkably, no one else in the fledgling industry had figured out this cause-and-effect; so no one else pursued working-to-gauge with Ford's near-religious zeal.

Ford also benefitted from recent advances in machine tools able to work on *prehardened* metals. The warping that occurred as machined parts were being hardened had been the bane of previous attempts to standardize parts. Once the warping problem was solved, Ford was able to develop innovative designs that reduced the number of parts needed and made these parts easy to attach. For example, Ford's four-cylinder engine block consisted of a single, complex casting. Competitors cast each cylinder separately and bolted the four together.

Taken together, interchangeability, simplicity, and ease of attachment gave Ford tremendous advantages over his competition. For one, he could eliminate the skilled fitters who had always formed the bulk of every assembler's labor force.

Ford's first efforts to assemble his cars, beginning in 1903, involved setting up assembly stands on which a whole car was built, often by one fitter. In 1908, on the eve of the introduction of the Model T, a Ford assembler's average task cycle—the amount of time he worked before repeating the same operations—totaled 514 minutes, or 8.56 hours. Each worker would assemble a large part of a car before moving on to the next. For example, a worker might put all the mechanical parts—wheels, springs, motor, transmission, generator—on the chassis, a set of activities that took a whole day to complete. The assembler/fitters performed the same set of activities over and over at their stationary assembly stands. They had to get the necessary parts, file them down so they would fit (Ford hadn't yet achieved perfect interchangeability of parts), then bolt them in place.

The first step Ford took to make this process more efficient was to deliver the parts to each work station. Now the assemblers could remain at the same spot all day.

Then, around 1908, when Ford finally achieved perfect part interchangeability, he decided that the assembler would perform only a single task and move from vehicle to vehicle around the assembly hall. By August of 1913, just before the moving assembly line was introduced, the task cycle for the average Ford assembler had been reduced from 514 to 2.3 minutes.

Naturally, this reduction spurred a remarkable increase in productivity, partly because complete familiarity with a single task meant the worker could perform it faster, but also because all filing and adjusting of parts had by now been eliminated. Workers simply popped on parts that fitted every time.

Ford's innovations must have meant huge savings over earlier production techniques, which required workers to file and fit each imperfect part. Unfortunately, the significance of this giant leap toward mass production went largely unappreciated, so we have no accurate estimates of the amount of effort—and money—that the minute division of labor and perfect interchangeability saved. We do know that it was substantial, probably much greater than the savings Ford realized in the next step, the introduction in 1913 of the continuous-flow assembly line.

Ford soon recognized the problem with moving the worker from assembly stand to assembly stand: Walking, even if only for a yard or two, took time, and jam-ups frequently resulted as faster workers overtook the slower workers in front of them. Ford's stroke of genius in the spring of 1913, at his new Highland Park plant in Detroit, was the introduction of the moving assembly line, which brought the car past the stationary worker. This innovation cut cycle time from 2.3 minutes to 1.19 minutes; the difference lay in the time saved in the worker's standing still rather than walking and in the faster work pace, which the moving line could enforce.

With this highly visible change, people finally began to pay attention, so we have well-documented accounts of the manufacturing effort this innovation saved. Journalists Horace Arnold and Fay Faurote, for example, writing in *Engineering Magazine* in 1915, compared the number of items assembled by the same number of workers using stationary and moving-assembly techniques and gave the world a vivid and dramatic picture of what Ford had wrought (see Figure 2.1)

Productivity improvements of this magnitude caught the attention and sparked the imagination of other auto assemblers. Ford, his competitors soon realized, had made a remarkable

# THE RISE AND FALL OF MASS PRODUCTION

**FIGURE 2.1**

**Craft Production versus Mass Production in the Assembly Hall: 1913 versus 1914**

| Minutes of Effort to Assemble: | Late Craft Production, Fall 1913 | Mass Production, Spring 1914 | Percent Reduction in Effort |
|---|---|---|---|
| Engine | 594 | 226 | 62 |
| Magneto | 20 | 5 | 75 |
| Axle | 150 | 26.5 | 83 |
| Major Components into a Complete Vehicle | 750 | 93 | 88 |

*Note:* "Late craft production" already contained many of the elements of mass production, in particular consistently interchangeable parts and a minute division of labor. The big change from 1913 to 1914 was the transition from stationary to moving assembly.

*Source:* Calculated by the authors from data given in David A. Hounshell, *From the American System to Mass Production, 1800–1932*, Baltimore: Johns Hopkins University Press, 1984, pp. 248, 254, 255, and 256. Hounshell's data are based on the observations of the journalists Horace Arnold and Fay Faurote as reported in their volume *Ford Methods and the Ford Shops*, New York: Engineering Magazine, 1915.

discovery. His new technology actually reduced capital requirements. That's because Ford spent practically nothing on his assembly line—less than $3,500 at Highland Park[6]—and it speeded up production so dramatically that the savings he could realize from reducing the inventory of parts waiting to be assembled far exceeded this trivial outlay.

(Ford's moving assembly consisted of two strips of metal plates—one under the wheels on each side of the car—that ran the length of the factory. At the end of the line, the strips, mounted on a belt, rolled under the floor and returned to the beginning. The device was quite similar to the long rubber belts that now serve as walkways in some airports. Since Ford needed only the belt and an electric motor to move it, his cost was minimal.)

Even more striking, Ford's discovery simultaneously reduced the amount of human effort needed to assemble an automobile. What's more, the more vehicles Ford produced, the more the cost per vehicle fell. Even when it was introduced in 1908, Ford's Model T, with its fully interchangeable parts, cost less than its rivals. By the time Ford reached peak production volume of 2 million identical vehicles a year in the early 1920s, he had cut the real cost to the consumer by an additional two-thirds.[7]

To appeal to his target market of average consumers, Ford had also designed unprecedented ease of operation and maintainability into his car. He assumed that his buyer would be a farmer with a modest tool kit and the kinds of mechanical skills needed for fixing farm machinery. So the Model T's owner's manual, which was written in question-and-answer form, explained in sixty-four pages how owners could use simple tools to solve any of the 140 problems likely to crop up with the car.

For example, owners could remove cylinder-head carbon, which causes knocking and power loss, from chamber roofs and piston crowns by loosening the fifteen cap screws that held the cylinder head and using a putty knife as a scraper. Similarly, a single paragraph and one diagram told customers how to remove carbon deposits from their car's valves with the Ford Valve Grinding Tool, which came with the auto.[8] And, if a part needed replacement, owners could buy a spare at a Ford dealer and simply screw or bolt it on. With the Ford Model T, there was no fitting required.

Ford's competitors were as amazed by this designed-in repairability as by the moving assembly line. This combination of competitive advantages catapulted Ford to the head of the world's motor industry and virtually eliminated craft-production companies unable to match its manufacturing economies. (As we pointed out earlier, however, a few European craft-based producers of ultra-low-volume luxury cars could ignore the juggernaut of mass production.)

Henry Ford's mass production drove the auto industry for more than half a century and was eventually adopted in almost every industrial activity in North America and Europe. Now, however, those same techniques, so ingrained in manufacturing philosophy, are thwarting the efforts of many Western companies to move ahead to lean production.

What precisely are the characteristics of mass production as pioneered by Ford in 1913 and persisting in so many companies today? Let's take a look.

### Work Force

Ford not only perfected the interchangeable part, he perfected the interchangeable worker. By 1915, when the assembly lines at

Highland Park were fully installed and output reached capacity, assembly workers numbered more than 7,000. Most were recent arrivals to Detroit, often coming directly from the farm. Many more were new to the United States.

A 1915 survey revealed that Highland Park workers spoke more than fifty languages and that many of them could barely speak English.[9] How could this army of strangers cooperate to produce a greater volume of a complex product (the Model T) than any company had previously imagined—and do it with consistent accuracy?

The answer lay in taking the idea of the division of labor to its ultimate extreme. The skilled fitter in Ford's craft-production plant of 1908 had gathered all the necessary parts, obtained tools from the tool room, repaired them if necessary, performed the complex fitting and assembly job for the entire vehicle, then checked over his work before sending the completed vehicle to the shipping department.

In stark contrast, the assembler on Ford's mass-production line had only one task—to put two nuts on two bolts or perhaps to attach one wheel to each car. He didn't order parts, procure his tools, repair his equipment, inspect for quality, or even understand what the workers on either side of him were doing. Rather, he kept his head down and thought about other things. The fact that he might not even speak the same language as his fellow assemblers or the foreman was irrelevant to the success of Ford's system. (Our use of "he," "him," and "his" is deliberate; until World War II, workers in auto factories in the United States and Europe were exclusively male.)

Someone, of course, did have to think about how all the parts came together and just what each assembler should do. This was the task for a newly created professional, the industrial engineer. Similarly, someone had to arrange for the delivery of parts to the line, usually a production engineer who designed conveyor belts or chutes to do the job. Housecleaning workers were sent around periodically to clean up work areas, and skilled repairmen circulated to refurbish the assemblers' tools. Yet another specialist checked quality. Work that was not done properly was not discovered until the end of the assembly line, where another group of workers was called into play—the rework men, who retained many of the fitters' skills.

With this separation of labor, the assembler required only a few minutes of training. Moreover, he was relentlessly disciplined by the pace of the line, which speeded up the slow and slowed down the speedy. The foreman—formerly the head of a whole area of the factory with wide-ranging duties and responsibilities, but now reduced to a semiskilled checker—could spot immediately any slacking off or failure to perform the assigned task. As a result, the workers on the line were as replaceable as the parts on the car.

In this atmosphere, Ford took it as a given that his workers wouldn't volunteer any information on operating conditions—for example, that a tool was malfunctioning—much less suggest ways to improve the process. These functions fell respectively to the foreman and the industrial engineer, who reported their findings and suggestions to higher levels of management for action. So were born the battalions of narrowly skilled indirect workers—the repairman, the quality inspector, the housekeeper, and the rework specialist, in addition to the foreman and the industrial engineer. These workers hardly existed in craft production. Indeed, Faurote and Arnold never thought to look for them when preparing the productivity figures shown in Figure 2.1.[10] These figures count only the direct workers standing on the assembly line. However, indirect workers became ever more prominent in Fordist, mass-production factories as the introduction of automation over the years gradually reduced the need for assemblers.

Ford was dividing labor not only in the factory, but also in the engineering shop. Industrial engineers took their places next to the manufacturing engineers who designed the critical production machinery. They were joined by product engineers, who designed and engineered the car itself. But these specialties were only the beginning.

Some industrial engineers specialized in assembly operations, others in the operation of the dedicated machines making individual parts. Some manufacturing engineers specialized in the design of assembly hardware, others designed the specific machines for each special part. Some product engineers specialized in engines, others in bodies, and still others in suspensions or electrical systems.

These original "knowledge workers"—individuals who manipulated ideas and information but rarely touched an actual car or even entered the factory—replaced the skilled machine-shop owners and the old-fashioned factory foremen of the earlier craft era. Those worker-managers had done it all—contracted with the assembler, designed the part, developed a machine to make it, and, in many cases, supervised the operation of the machine in the workshop. The fundamental mission of these new specialists, by contrast, was to design tasks, parts, and tools that could be handled by the unskilled workers who made up the bulk of the new motor-vehicle industry work force.

In this new system, the shop-floor worker had no career path, except perhaps to foreman. But the newly emerging professional engineers had a direct climb up the career ladder. Unlike the skilled craftsman, however, their career paths didn't lead toward ownership of a business. Nor did they lie within a single company, as Ford probably hoped. Rather, they would advance within their profession—from young engineer-trainee to senior engineer, who, by now possessing the entire body of knowledge of the profession, was in charge of coordinating engineers at lower levels.

Reaching the pinnacle of the engineering profession often meant hopping from company to company over the course of one's working life. As time went on and engineering branched into more and more subspecialties, these engineering professionals found they had more and more to say to their subspecialists and less and less to say to engineers with other expertise. As cars and trucks became ever more complicated, this minute division of labor within engineering would result in massive dysfunctions, which we'll look at in Chapter 5.

## Organization

Henry Ford was still very much an assembler when he opened Highland Park. He bought his engines and chassis from the Dodge brothers, then added a host of items ordered from other firms to make a complete vehicle. By 1915, however, Ford had taken all these functions in-house and was well on his way to achieving complete vertical integration (that is, making everything connected with his cars from the basic raw materials on up). This development reached its logical conclusion in the Rouge complex in Detroit, which opened in 1931. Ford pursued vertical integration partly because he had perfected mass-production techniques before his suppliers had and could achieve substantial cost savings by doing everything himself. But he also had some other reasons: For one, his peculiar character caused him to profoundly distrust everyone but himself.

However, his most important reason for bringing everything in-house was the fact that he needed parts with closer tolerances and on tighter delivery schedules than anyone had previously imagined. Relying on arm's-length purchases in the open marketplace, he figured, would be fraught with difficulties. So he decided to replace the mechanism of the market with the "visible hand" of organizational coordination.

Alfred Chandler, a professor at the Harvard Business School, coined the term "visible hand" in 1977. In his book of the same title, he attempted to provide a defense for the modern large firm.[11] Proponents of Adam Smith's "invisible hand" theory (which argued that if everyone pursued his or her own self-interest, the free market would of itself produce the best outcome for society as a whole) were disturbed by the rise in the twentieth century of the vertically integrated modern corporation. In their view, vertical integration interfered with free-market forces. Chandler argued that a visible hand was critical if modern corporations were to introduce necessary predictability into their operations.

Chandler used the term simply to mean obtaining needed raw materials, services, and so forth from internal operating divisions coordinated by senior executives at corporate headquarters. The invisible hand, by contrast, meant buying necessary parts and services from independent firms with no financial or other relationship to the buyer. Transactions would be based on price, delivery time, and quality, with no expectation of any long-term or continuing relationship between the buyer and the seller. The problem, as we will see, was that total vertical integration introduced bureaucracy on such a vast scale that it brought its own problems, with no obvious solutions.

The scale of production possible—and necessary—with Ford's system led to a second organizational difficulty, this time caused by shipping problems and trade barriers. Ford wanted to produce the entire car in one place and sell it to the whole world. But the shipping systems of the day were unable to transport huge volumes of finished automobiles economically without damaging them.

Also, government policies, then as now, often imposed trade barriers on finished units. So Ford decided to design, engineer, and produce his parts centrally in Detroit. The cars, however, would be assembled in remote locations. By 1926, Ford automobiles were assembled in more than thirty-six cities in the United States and in nineteen foreign countries.[12]

It wasn't long before this solution created yet another problem: One standard product just wasn't suited to all world markets. For example, to Americans, Ford's Model T seemed like a small car, particularly after the East Texas oil discoveries pushed gasoline prices down and made longer travel by car economically feasible. However, in England and in other European countries, with their crowded cities and narrow roads, the Model T seemed much larger. In addition, when the Europeans failed to find any oil at home, they began to tax gasoline heavily in the 1920s to reduce imports. The Europeans soon began to clamor for a smaller car than Ford wanted to supply.

Moreover, massive direct investment in foreign countries created resentment of Ford's dominance of local industry. In England, for example, where Ford had become the leading auto manufacturer by 1915, his pacifism in World War I was roundly denounced, and the company's local English managers finally convinced Detroit to sell a large minority stake in the business to Englishmen to diffuse hostility. Ford encountered barriers in Germany and France as well after World War I, as tariffs were steadily raised on parts and complete vehicles. As a result, by the early 1930s, Ford had established three fully integrated manufacturing systems in England, Germany, and France. These companies produced special products for national tastes and were run by native managers who tried to minimize meddling from Detroit.

## Tools

The key to interchangeable parts, as we saw, lay in designing new tools that could cut hardened metal and stamp sheet steel with absolute precision. But the key to *inexpensive* interchangeable parts would be found in tools that could do this job at high volume with low or no set-up costs between pieces. That is, for a machine to do something to a piece of metal, someone must put the metal in the machine, then someone may need to adjust the machine. In the craft-production system—where a single machine could do many tasks but required lots of adjustment—this was the skilled machinist's job.

Ford dramatically reduced set-up time by making machines that could do only one task at a time. Then his engineers perfected simple jigs and fixtures for holding the work piece in this dedicated machine. The unskilled workers could simply snap the piece in place and push a button or pull a lever for the machine to perform the required task. This meant the machine could be loaded and unloaded by an employee with five minutes' training. (Indeed, loading Ford's machines was exactly like assembling parts in the assembly line: The parts would fit only one way, and the worker just popped them on.)

In addition, because Ford made only one product, he could place his machines in a sequence so that each manufacturing step led immediately to the next. Many visitors to Highland Park felt that Ford's factory was really one vast machine with each production step tightly linked to the next. Because set-up times were reduced from minutes—or even hours—to seconds, Ford could get much higher volume from the same number of machines. Even more important, the engineers also found a way to machine many parts at once. The only penalty with this system was inflexibility. Changing these dedicated machines to do a new task was time-consuming and expensive.

Ford's engine-block milling machine is a good example of his new system. In almost every auto engine, then and now, the top of the engine block is mated to the bottom of the cylinder head to form a complete engine. To maintain compression in the cylinders, the fit between block and head must be absolutely flush. So the top of the block and the bottom of the cylinder head has to be milled with a grinding tool.

At Henry Leland's Cadillac plant in Detroit (where, incidentally, consistent interchangeability for all the parts in a motor vehicle was achieved for the first time in 1906), a worker would load each block in a milling machine, then carefully mill it to specification. The worker would repeat the process for the cylinder heads, which were loaded one at a time into the same milling machine.

In this way, the parts were interchangeable, the fit between block and head was flush, and the milling machine could work on a wide variety of parts. But this process had a down side: the time and effort—and therefore expense—it took for the skilled machinist who operated the machine.

At Highland Park in 1915, Ford introduced two dedicated machines, one for milling blocks and the other for milling heads—

not just one at a time, but fifteen at a time for blocks and thirty at a time for heads. Even more significant, a fixture on both machines allowed unskilled workers to snap the blocks and heads in place on a side tray, while the previous lot was being milled. The worker then pushed the whole tray into the miller, and the process proceeded automatically. Now the entire skill in milling was embodied in the machine, and the cost of the process plummeted.

Ford's tools were highly accurate and in many cases automated or nearly so, but they were also dedicated to producing a single item, in some cases to an absurd degree. For example, Ford purchased stamping presses, used to make sheet-steel parts, with die spaces large enough to handle only a specific part. When the factory needed a larger part because of a specification change or, in 1927, for the completely redesigned Model A, Ford often discarded the machinery along with the old part or model.

## Product

Ford's original mass-produced product, the Model T, came in nine body styles—including a two-seat roadster, a four-seat touring car, a four-seat covered sedan, and a two-seat truck with a cargo box in the rear. However, all rode on the same chassis, which contained all the mechanical parts. In 1923, the peak year of Model T production, Ford produced 2.1 million Model T chassis, a figure that would prove to be the high-water mark for standardized mass production (although the VW Beetle later equaled it).

The success of his automobiles was based first and foremost on low prices, which kept falling. Ford dropped his prices steadily from the day the Model T was introduced. Some of the reduction had to do with shifts in general consumer prices—before governments tried to stabilize the economy, consumer prices went down as well as up—but mostly it was a matter of growing volume permitting lower costs that, in turn, generated higher volume. At the end of its run in 1927, however, Ford was facing falling demand for the Model T and was undoubtedly selling below cost. (Demand fell, because General Motors was offering a more modern product for only a little more money. Moreover, a one-year-old GM automobile was less expensive than a new Ford.)

The Ford car's amazing popularity also stemmed from its durability of design and materials and, as noted, from the fact that the average user could easily repair it. Concerns that buyers rank highest today hardly existed in Ford's world.

For example, fits and finishes—or the cosmetic aspects of a car, such as the fender panels coming together without gaps, a lack of dribbles in the paint, or the doors making a satisfying clunking sound when slammed—weren't a concern for Ford's customers. The Model T had no exterior sheet metal except the hood; the paint was so crude that you would hardly have noticed dribbles; and several of the body styles had no doors at all.

As for breakdowns or problems in daily use—engines that stumble, say, or mysterious electrical difficulties, such as the "check engine" signal that comes on periodically in some cars—these, too, didn't bother Ford's buyers. If the Model T engine stumbled, they simply looked for the cause in the question-and-answer booklet the company provided and fixed the problem. For example, they might drain the gas tank and pour the fuel back through a chamois to strain any water out. The bottom line: If a part didn't fit properly or was installed slightly out of tolerance, the owner was expected to fix it. And, since all cars broke down frequently, ease of repair was key.

At Highland Park, Ford rarely inspected finished automobiles. No one ran an engine until the car was ready to drive away from the end of the assembly line, and no Model T was ever road-tested.

Nonetheless, despite a manufacturing system that probably did not deliver very high quality in our modern sense, Ford was able to dominate what soon became the world's largest industry by becoming the first to master the principles of mass production. It wasn't until fifty years later that plants organized on lean-production principles could deliver near-perfect quality without extensive end-of-the-line inspections and large amounts of rework.

## THE LOGICAL LIMITS OF MASS PRODUCTION: THE ROUGE

True mass production began with Highland Park, but the end wasn't yet in sight. Ford believed that the last piece in the puzzle was to apply a "visible hand" to every step in production, from raw materials to finished vehicle. This he attempted to do at the Rouge complex, which he opened near Detroit in 1927. Smaller-scale duplicates of the Rouge were opened at Dagenham, England, and Cologne, Germany, in 1931.

At these facilities, Ford continued his obsession with a single product—the Model A at the Rouge, the Model Y at Dagenham, and the Ford V8 in Germany. He also added a steel mill and a glass factory to the metal-forming and -cutting activities that took place at Highland Park. All the necessary raw materials now came in one gate, while finished cars went out the other gate. Ford had succeeded in completely eliminating the need for outside assistance.

He even added raw materials and transport to the visible hand—through a wholly owned rubber plantation in Brazil, iron mines in Minnesota, Ford ships to carry iron ore and coal through the Great Lakes to the Rouge, and a railroad to connect Ford production facilities in the Detroit region.

In the end, Ford attempted to mass-produce everything—from food (through tractor manufacture and a soybean extraction plant) to air transportation (by means of the Ford TriMotor, which was supposed to reduce the price of commercial air traffic, and the Ford "Flying Fliver," which was intended as the airborne equivalent of the Model T). Ford's idea was that by making everything, from food to tractors to airplanes, in a standardized form at high volume, he could dramatically reduce the cost of products and make the masses rich. He financed all his projects internally, for Ford loathed banks and outside investors and was determined to maintain total control of his company.

Eventually, these steps beyond Highland Park all came to naught, partly because the synergy among industries, which industrialists repeatedly seek and seldom find, was never there, but also because Ford himself had absolutely no idea how to organize a global business except by centralizing all decision-making in the one person at the top—himself. This concept was unworkable even when Ford was in his prime, and it nearly drove the company under when his mental powers declined in the 1930s.

## SLOAN AS A NECESSARY COMPLEMENT TO FORD

Alfred Sloan at General Motors already had a better idea in the early 1920s when he was called in to straighten out the messes that William Durant, General Motors' mercurial founder, had made. Durant was the classic empire-building financier. He had absolutely no idea how to manage anything once he bought it. He therefore wound up with a dozen car companies, each managed separately with a high degree of product overlap. Because he had no way to know what was going on in these companies, beyond quarterly profit-and-loss statements, he was repeatedly surprised to discover that too many cars were being manufactured for market conditions or that not enough raw materials were available to sustain production. A burst of overproduction heading into the economic slump of 1920 finally did him in; his bankers insisted that someone with management skills take the helm. So Pierre du Pont, chairman of E. I. du Pont, became chairman of General Motors and, in turn, made Sloan GM's president.

An MIT graduate (he contributed a block of his GM earnings to found the Sloan School of Management at MIT after World War II), Sloan gained control in the early 1900s of the Hyatt Roller Bearing Company, a firm purchased by Billy Durant around 1915. He was vice-president of GM when Durant was ousted; he gained the presidency on the basis of a memo he wrote in 1919 on how to run a multidivisional company.

Sloan quickly saw that GM had two critical problems to solve if it was going to succeed at mass production and oust Ford as the industry leader: The company had to manage professionally the enormous enterprises that the new production techniques had both necessitated and made possible, and it had to elaborate on Ford's basic product so it would serve, as Sloan put it, "every purse and purpose."

Ford Motor Company, of course, didn't suffer from GM's product overlap problem, because Ford produced only one product. It did, however, have all the organizational problems, but Henry Ford refused to acknowledge them. He succeeded with mass production in the factory, but he could never devise the organization and management system he needed to manage effectively the total system of factories, engineering operations, and marketing systems that mass production called for. Sloan would make the system Ford had pioneered complete, and it is this complete system to which the term *mass production* applies today.

Sloan swiftly found a solution for each of GM's difficulties. To resolve the management problem he created decentralized divisions managed objectively "by the numbers" from a small corporate headquarters. That is, Sloan and the other senior executives oversaw each of the company's separate profit centers—the five car divisions and the divisions making components such as generators (Delco), steering gears (Saginaw), and carburetors (Rochester). Sloan and his executive group demanded detailed reports at frequent intervals on sales, market share, inventories, and profit and loss and reviewed capital budgets when the divisions required funds from the central corporate coffer.

Sloan thought it both unnecessary and inappropriate for senior managers at the corporate level to know much about the details of operating each division. If the numbers showed that performance was poor, it was time to change the general manager. General managers showing consistently good numbers were candidates for promotion to the vice-presidential level at headquarters.

To satisfy the broad market General Motors wanted to serve, Sloan developed a five-model product range that ran incrementally from cheap to expensive, from Chevrolet to Cadillac. It would, Sloan reasoned, fully accommodate potential buyers of every income throughout their lives.

Sloan had worked out this strategic solution to the company's problems by about 1925, although he only codified it for the world outside General Motors when he got around to writing his memoirs as he approached ninety in the 1960s.[13]

He also worked out solutions to two other major problems confronting the company. Through his links with DuPont and the Morgan Bank, he developed stable sources of outside funding, which would be available when needed.

Also, his idea of decentralized divisions domestically worked just as well in organizing and managing GM's foreign subsidiaries. Manufacturing and sales operations in Germany, Britain, and many other countries became self-reliant companies managed by the numbers from Detroit. The arrangement demanded very little management time or direct supervision.

It isn't giving Sloan too much credit to say that his basic management ideas solved the last pressing problems inhibiting the spread of mass production. New professions of financial managers and marketing specialists were created to complement the engineering professions, so that every functional area of the firm now had its dedicated experts. The division of professional labor was complete.

Sloan's innovative thinking also seemed to resolve the conflict between the need for standardization to cut manufacturing costs and the model diversity required by the huge range of consumer demand. He achieved both goals by standardizing many mechanical items, such as pumps and generators, across the company's entire product range and by producing these over many years with dedicated production tools. At the same time, he annually altered the external appearance of each car and introduced an endless series of "hang-on features," such as automatic transmissions, air conditioning, and radios, which could be installed in existing body designs to sustain consumer interest.

Sloan's innovations were a revolution in marketing and management for the auto industry. However, they did nothing to change the idea, institutionalized first by Henry Ford, that the workers on the shop floor were simply interchangeable parts of the production system. So, on the shop floor matters went from bad to much worse.

Ford himself was happy enough with the high rates of turnover his labor philosophy and practices encouraged. Nonetheless, he realized that once the continuous-flow system was fully in place at Highland Park in 1914, his company's efficiency was so much higher than its rivals that he could afford simultaneously to double wages (to the famous five-dollar day) and dramatically slash prices. These actions permitted him to portray himself as a paternalistic employer (and avoid unions), while he drove his craft-based competitors to the wall.

The trouble with the higher wage, as it turned out, was that it worked: Turnover slowed as Ford's workers decided to stay in their jobs. Eventually they began to stop dreaming about a return to the farm or to the old country and realize that a job at Ford was likely to be their life's work. When that realization dawned, their conditions of employment rapidly came to seem less and less bearable.

Furthermore, the auto market turned out to be even more cyclical than the rest of the economy. American car companies, of course, considered their work force a variable cost, and they turned workers away from their plants at the first sign of a downturn in sales. All this meant that by the time of the Great Depression the conditions for a successful union movement in the auto industry were fully in place.

This was, however, a mass-production union movement. Its leadership fully accepted both the role of management and the inherent nature of work in an assembly-line factory. Not surprisingly, then, when the United Auto Workers finally signed agreements with what had become the Big Three in the late 1930s, the main issues were seniority and job rights; the movement was called job-control unionism.[14]

The cyclical nature of the industry meant that some workers would be laid off frequently, so seniority—not competence—became the key determinant of who would go and who would stay. And because some jobs were easier (or more interesting) than others but all paid roughly the same wage, seniority also became the principle that governed job assignments as well. The result was an ever-growing list of work rules that unquestionably reduced the efficiency of Ford's mass-production factory as workers fought continually for equity and fairness.

## THE HEYDAY OF MASS PRODUCTION: AMERICA IN 1955

Take Ford's factory practices, add Sloan's marketing and management techniques, and mix in organized labor's new role in controlling job assignments and work tasks, and you have mass production in its final mature form. For decades this system marched from victory to victory. The U.S. car companies dominated the world automotive industry, and the U.S. market accounted for the largest percentage of the world's auto sales. Companies in practically every other industry adopted similar methods, usually leaving a few craft firms in low-volume niches.

As no year had before, 1955 illustrated just how large and pervasive the auto industry and the system on which it was based had become. This marked the first year that more than 7 million automobiles were sold in the United States. It was also the year in which Sloan retired after thirty-four years as either president or chairman of General Motors.

Three giant enterprises—Ford, GM, and Chrysler—accounted for 95 percent of all sales, and six models accounted for 80 percent of all cars sold. All vestiges of craft production, once the way of all industry, were now gone in the United States.

Glory is fleeting, however, as the then mighty U.S. auto industry has now learned. Ironically, 1955 was also the year that the downhill slide began, as Figures 2.2 and 2.3 show. The share of market claimed by imports began its steady rise. Their early perfection of mass production could no longer sustain these U.S. companies in their leading positions.

**FIGURE 2.2**

**Shares of World Motor Vehicle Production by Region, 1955–1989**

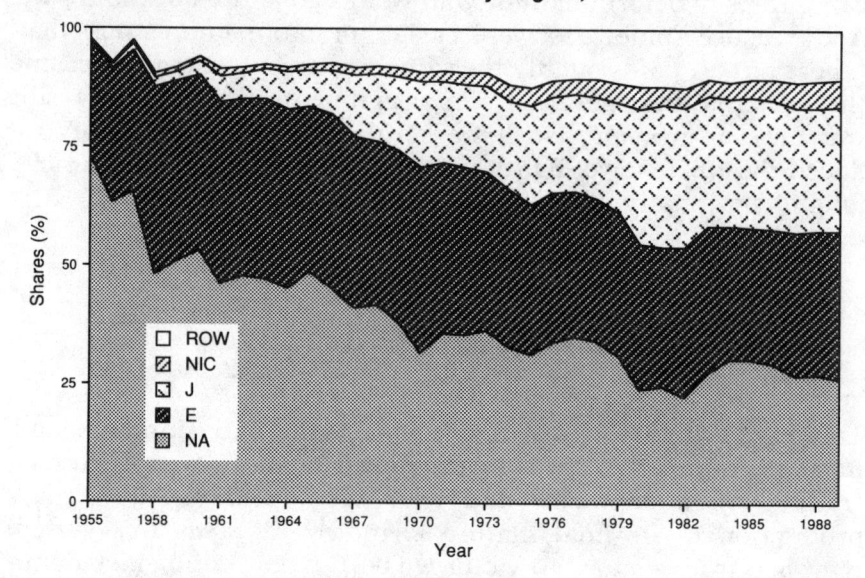

*Note:* This figure includes all vehicles produced within the three major regions, by all companies operating in those regions. In addition, it groups the production of the newly industrializing countries and of the rest of the world.

NA = North America: United States and Canada
E = Western Europe, including Scandinavia
J = Japan
NIC = Newly industrializing countries, principally Korea, Brazil, and Mexico
ROW = Rest of the world, including the Soviet Union, Eastern Europe, and China

*Source:* Calculated by the authors from *Automotive News Market Data Book*, 1990 edition, p. 3.

## THE DIFFUSION OF MASS PRODUCTION

A major reason the Big Three American firms were losing their competitive advantage was that by 1955 mass production had become commonplace in countries across the world. Many people, in fact, had expected the American lead to narrow much earlier, in the years immediately after World War I. Even before the war, a steady stream of pilgrims, including André Citroën, Louis Renault, Giovanni Agnelli (of Fiat), Herbert Austin, and William Morris (of Morris and MG in England), had visited Highland Park. Henry Ford was remarkably open in discussing his techniques

## FIGURE 2.3

**Share of the American Car Market Held by the American-Owned Companies, 1955–1989**

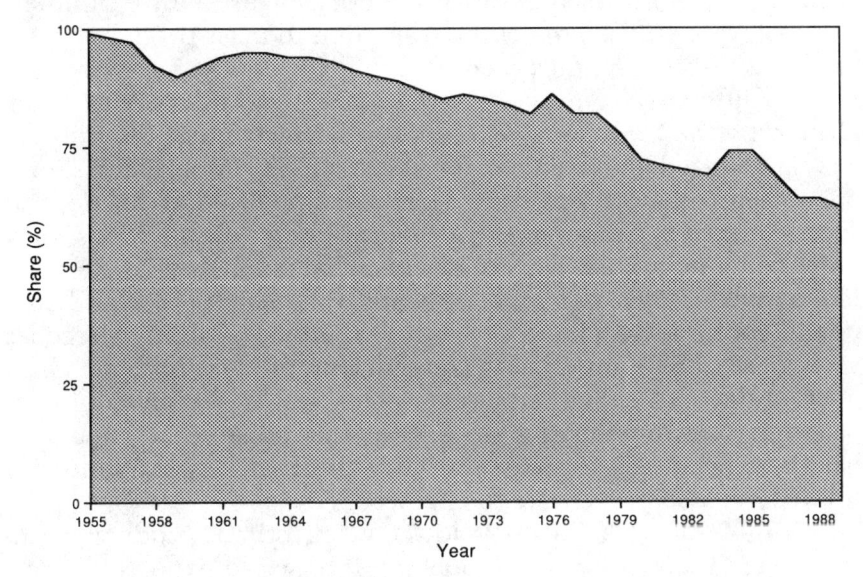

*Note:* These shares include vehicles imported by the American-owned firms from their wholly owned and joint-venture factories abroad. They do not include "captive" imports purchased from independent foreign firms.

*Source:* 1955–1981 from *Automotive News Market Data Book*, based on vehicle registrations.
1982–1989 from *Ward's Automotive Reports*, based on vehicle sales.

with them, and, in the 1930s, he directly demonstrated every aspect of mass production in Europe with his Dagenham and Cologne factories.

The basic ideas underlying mass production had, therefore, been freely available in Europe for years before the onset of World War II. However, the economic chaos and narrow nationalism existing there during the 1920s and early 1930s, along with a strong attachment to the craft-production traditions, prevented them from spreading very far. At the end of the 1930s, Volkswagen and Fiat began ambitious plans for mass production at Wolfsburg and Mirafiori, but World War II soon put civilian production on hold.

So, it wasn't until the 1950s, more then thirty years after Henry Ford pioneered high-volume mass production, that this technology, unremarkably commonplace in the United States, fully diffused beyond Ford's native turf. By the late 1950s, Wolfsburg (VW), Flins (Renault), and Mirafiori (Fiat) were producing at a scale comparable to Detroit's major facilities. Furthermore, a number of the European craft-production firms, led by Daimler-Benz (Mercedes), also made the transition to mass production.

All these companies were offering products that were distinctly different from the standard-size car and pickup truck favored by the U.S. manufacturers. In the early days, the Europeans specialized in two types of cars that the Americans didn't offer: compact, economy cars, exemplified by the VW Beetle, and sporty, fun-to-drive cars, such as the MG. Later, in the 1970s, they redefined the luxury car as a somewhat smaller vehicle with higher technology and more sporting road manners (the 3,500-pound, fuel-injected, independently suspended, unibody Mercedes versus the 5,000-pound, carbureted, straight-axle, body-on-chassis Cadillac). (The unibody car weighs less for a given size of passenger compartment than a body-on-chassis car. Though it has the advantages of greater rigidity and thus less of a tendency to rattle, it also costs more to engineer.)

Combined with Europe's lower wages, these product variations were their competitive opening into world export markets. And, like the Americans before them, Europeans racked up success after success in foreign markets over a period of twenty-five years, from the early 1950s into the 1970s.

They also concentrated—as Detroit did not during this time—on introducing new product features. European innovations in the 1960s and 1970s included front-wheel drive, disc brakes, fuel injection, unitized bodies, five-speed transmissions, and engines with high power-to-weight ratios. (Unitized bodies have no frame of steel beams under the car. Instead, like a tin can, the surface sheet metal holds the car together.) The Americans, by contrast, were the leaders in comfort features—air conditioning, power steering, stereos, automatic transmissions, and massive (and very smooth) engines.

History could have gone the Americans' way if fuel prices had continued to fall—as they did for a generation, up until 1973—and if Americans had continued to demand cars that isolated them from their driving environment. However, energy prices soared and younger Americans, particularly those with money, wanted something fun to drive. Detroit's problem was that its "hang-on" features, such as air conditioning and stereos, could easily be added to existing European cars. But it would take a total redesign of the American vehicles and new production tools to introduce more space-efficient bodies, more responsive suspensions, and more fuel-efficient engines.

However, as became apparent in the late 1980s and as we will show in the following chapters, the European production systems were nothing more than copies of Detroit's, but with less efficiency and accuracy in the factory.

European auto plants experienced in the 1950s what the Americans had experienced in the 1930s. During the early postwar years, most European plants employed large numbers of immigrants—Turks and Yugoslavs in Germany, Sicilians and other southern Italians in Italy, and Moroccans and Algerians in France—in the interchangeable assembler jobs.

Some of these people returned home as the postwar European labor shortage eased. Others, however, stayed, to be joined by larger numbers of native workers. Eventually, just as had happened in the United States, the workers in Turin, Paris, and Wolfsburg realized that mass-production work was not a way station to self-employment back home; it was, instead, their life's work. Suddenly the interchangeable, dead-end monotony of mass-production plants began to seem unbearable. A wave of unrest followed.

The European mass-production systems were patched up in the 1970s by increasing wages and steadily decreasing the weekly hours of work. European car makers conducted a few marginal experiments as well with worker participation, such as the one at Volvo's Kalmar plant, which—in a revival of Henry Ford's assembly hall of 1910—reintroduced craft techniques by giving small groups of workers responsibility for assembling a whole vehicle. In addition, the sobering economic conditions after 1973 damped worker expectations and reduced employment alternatives.

These were only palliatives, however. In the 1980s, European workers continued to find mass-production work so unrewarding that the first priority in negotiations continued to be reducing hours spent in the plant.

This situation of stagnant mass production in both the United States and Europe might have continued indefinitely if a new motor industry had not emerged in Japan. The true significance of this industry was that it was not simply another replication of the by now venerable American approach to mass production. The Japanese were developing an entirely new way of making things, which we call *lean production*.

# THE RISE OF LEAN PRODUCTION

## 3

In the spring of 1950, a young Japanese engineer, Eiji Toyoda, set out on a three-month pilgrimage to Ford's Rouge plant in Detroit. In fact, the trip marked a second pilgrimage for the family, since Eiji's uncle, Kiichiro, had visited Ford in 1929.

Since that earlier time much had happened to the Toyoda family and the Toyota Motor Company they had founded in 1937.[1] (The founding family's name, Toyoda, means "abundant rice field" in Japanese, so marketing considerations called for a new name for the fledgling company. Accordingly, in 1936, the company held a public contest, which drew 27,000 suggestions. "Toyota," which has no meaning in Japanese, was the winner.)

Most of these events had been disastrous for the company: They had been thwarted by the military government in their effort to build passenger cars in the 1930s, and had instead made trucks, largely with craft methods, in the ill-fated war effort.

And, at the end of 1949, a collapse in sales forced Toyota to terminate a large part of the work force, but only after a lengthy strike that didn't end until Kiichiro resigned from the company to accept responsibility for management failures. In thirteen years of effort, the Toyota Motor Company had, by 1950, produced 2,685 automobiles, compared with the 7,000 the Rouge was pouring out in a single day.[2]

This was soon to change.

# THE RISE OF LEAN PRODUCTION

Eiji was not an average engineer, either in ability or ambition. After carefully studying every inch of the vast Rouge, then the largest and most efficient manufacturing facility in the world, Eiji wrote back to headquarters that he "thought there were some possibilities to improve the production system."[3]

But simply copying and improving the Rouge proved to be hard work. Back at home in Nagoya, Eiji Toyoda and his production genius, Taiichi Ohno, soon concluded—for reasons we will explain shortly—that mass production could never work in Japan. From this tentative beginning were born what Toyota came to call the Toyota Production System and, ultimately, lean production.[4]

## THE BIRTHPLACE OF LEAN PRODUCTION

Toyota is often called the most Japanese of the Japanese auto companies, being located in insular Nagoya rather than cosmopolitan Tokyo.[5] For many years its work force was composed largely of former agricultural workers. In Tokyo, the firm was often derided as "a bunch of farmers." Yet today, Toyota is regarded by most industry observers as the most efficient and highest-quality producer of motor vehicles in the world.

The founding Toyoda family succeeded first in the textile machinery business during the late nineteenth century by developing superior technical features on its looms. In the late 1930s, at the government's urging, the company entered the motor vehicle industry, specializing in trucks for the military. It had barely gone beyond building a few prototype cars with craft methods before war broke out and auto production ended. After the war, Toyota was determined to go into full-scale car and commercial truck manufacturing, but it faced a host of problems.

- The domestic market was tiny and demanded a wide range of vehicles—luxury cars for government officials, large trucks to carry goods to market, small trucks for Japan's small farmers, and small cars suitable for Japan's crowded cities and high energy prices.
- The native Japanese work force, as Toyota and other firms soon learned, was no longer willing to be treated as a variable cost or as interchangeable parts. What was more, the new labor laws introduced by the American occupation

greatly strengthened the position of workers in negotiating more favorable conditions of employment. Management's right to lay off employees was severely restricted, and the bargaining position of company unions representing all employees was greatly reinforced. The company unions used their strength to represent everyone, eliminating the distinction between blue- and white-collar workers, and secured a share of company profits in the form of bonus payments in addition to basic pay.[6]

Furthermore, in Japan there were no "guest workers"—that is, temporary immigrants willing to put up with substandard working conditions in return for high pay—or minorities with limited occupational choice.[7] In the West, by contrast, these individuals had formed the core of the work force in most mass-production companies.

- The war-ravaged Japanese economy was starved for capital and for foreign exchange, meaning that massive purchases of the latest Western production technology were quite impossible.
- The outside world was full of huge motor-vehicle producers who were anxious to establish operations in Japan and ready to defend their established markets against Japanese exports.

This last difficulty provoked a response from the Japanese government, which soon issued a prohibition on direct foreign investment in the Japanese motor industry. This prohibition was critical for Toyota (as well as other entrants in the Japanese auto industry) to gain a toehold in the car-making business. It wasn't enough, however, to guarantee the company's success beyond Japan.

Besides, the government nearly went too far. After the prohibition on foreign ownership and the imposition of high tariff barriers had encouraged a host of Japanese firms to enter the auto industry by the early 1950s, the Japanese Ministry of International Trade and Industry (MITI) had second thoughts. MITI believed that the first requirement of an internationally competitive auto industry was high production scale, so it proposed a series of plans to merge Japan's twelve embryonic car companies into a Japanese Big Two or Big Three to battle Detroit's Big Three. The merged companies were to specialize in different sizes of cars

# THE RISE OF LEAN PRODUCTION

to prevent "excessive" domestic competition and to gain high scale to compete on price in export markets.

What if these plans had succeeded?

The Japanese industry might have grown rapidly at first, but it would probably have shared the fate of the current Korean motor industry. That is, as the advantage of lower wages gradually disappeared, the new-entrant Japanese producers, with nothing new to offer in production techniques and limited competition at home, would have become also-rans in the world motor industry. They might have been able to protect their domestic market, but they would have posed no long-term threat to the established firms elsewhere in the world using the same techniques.

Instead, Toyota, Nissan, and the other companies defied MITI and set out to become full-range car producers with a variety of new models. Toyota's chief production engineer, Taiichi Ohno, quickly realized that employing Detroit's tools—and Detroit's methods—was not suited to this strategy. Craft-production methods were a well-known alternative but seemed to lead nowhere for a company intent on producing mass-market products. Ohno knew he needed a new approach, and he found it. We can look at the stamping shop for a good example of how his new techniques worked.[8]

## LEAN PRODUCTION: A CONCRETE EXAMPLE

More than sixty years have passed since the introduction of Henry Ford's Model A with its all-steel body. Yet, across the world, nearly all motor-vehicle bodies are still produced by welding together about 300 metal parts stamped from sheet steel.

Auto makers have produced these "stampings" by employing one of two different methods. A few tiny craft producers, such as Aston Martin, cut sheets of metal—usually aluminum—to a gross shape, then beat these blanks by hand on a die to their final shape. (A die is simply a hard piece of metal in the precise shape the sheet metal should assume under pounding.)

Any producer making more than a few hundred cars a year—a category that includes auto makers ranging from Porsche to General Motors—starts with a large roll of sheet steel. They run this sheet through an automated "blanking" press to produce a stack of flat blanks slightly larger than the final part they want.

They then insert the blanks in massive stamping presses containing matched upper and lower dies. When these dies are pushed together under thousands of pounds of pressure, the two-dimensional blank takes the three-dimensional shape of a car fender or a truck door as it moves through a series of presses.

The problem with this second method, from Ohno's perspective, was the minimum scale required for economical operation. The massive and expensive Western press lines were designed to operate at about twelve strokes per minute, three shifts a day, to make a million or more of a given part in a year. Yet, in the early days, Toyota's entire production was a few thousand vehicles a year.

The dies could be changed so that the same press line could make many parts, but doing so presented major difficulties. The dies weighed many tons each, and workers had to align them in the press with absolute precision. A slight misalignment produced wrinkled parts. A more serious misalignment could produce a nightmare in which the sheet metal melted in the die, necessitating extremely expensive and time-consuming repairs.

To avoid these problems, Detroit, Wolfsburg, Flins, and Mirafiori assigned die changes to specialists. Die changes were undertaken methodically and typically required a full day to go from the last part with the old dies to the first acceptable part from the new dies. As volume in the Western industry soared after World War II, the industry found an even better solution to the die-change problem. Manufacturers found they often could "dedicate" a set of presses to a specific part and stamp these parts for months, or even years, without changing dies.

To Ohno, however, this solution was no solution at all. The dominant Western practice required hundreds of stamping presses to make all the parts in car and truck bodies, while Ohno's capital budget dictated that practically the entire car be stamped from a few press lines.

His idea was to develop simple die-change techniques and to change dies frequently—every two to three hours versus two to three months—using rollers to move dies in and out of position and simple adjustment mechanisms.[9] Because the new techniques were easy to master and production workers were idle during the die changes, Ohno hit upon the idea of letting the production workers perform the die changes as well.

By purchasing a few used American presses and endlessly experimenting from the late 1940s onward, Ohno eventually perfected his technique for quick changes. By the late 1950s, he had reduced the time required to change dies from a day to an astonishing three minutes and eliminated the need for die-change specialists. In the process, he made an unexpected discovery—it actually cost less per part to make small batches of stampings than to run off enormous lots.

There were two reasons for this phenomenon. Making small batches eliminated the carrying cost of the huge inventories of finished parts that mass-production systems required. Even more important, making only a few parts before assembling them into a car caused stamping mistakes to show up almost instantly.

The consequences of this latter discovery were enormous. It made those in the stamping shop much more concerned about quality, and it eliminated the waste of large numbers of defective parts—which had to be repaired at great expense, or even discarded—that were discovered only long after manufacture. But to make this system work at all—a system that ideally produced two hours or less of inventory—Ohno needed both an extremely skilled and a highly motivated work force.

If workers failed to anticipate problems before they occurred and didn't take the initiative to devise solutions, the work of the whole factory could easily come to a halt. Holding back knowledge and effort—repeatedly noted by industrial sociologists as a salient feature of all mass-production systems—would swiftly lead to disaster in Ohno's factory.

## LEAN PRODUCTION: COMPANY AS COMMUNITY

As it happened, Ohno's work force acted to solve this problem for him in the late 1940s. Because of macroeconomic problems in Japan—the occupying Americans had decided to stamp out inflation through credit restrictions, but overdid it and caused a depression instead—Toyota found its nascent car business in a deep slump and was rapidly exhausting loans from its bankers.

The founding family, led by President Kiichiro Toyoda, proposed—as a solution to the crisis—firing a quarter of the work force. However, the company quickly found itself in the midst of a revolt that ultimately led to its workers occupying the factory. Moreover, the company's union was in a strong position to win the strike. In 1946, when the Japanese government, under American prompting, strengthened the rights of unions, including management, and then imposed severe restrictions on the ability of company owners to fire workers, the balance of power had shifted to the employees.

After protracted negotiations, the family and the union worked out a compromise that today remains the formula for labor relations in the Japanese auto industry. A quarter of the work force was terminated as originally proposed. But Kiichiro Toyoda resigned as president to take responsibility for the company's failure, and the remaining employees received two guarantees. One was for lifetime employment; the other was for pay steeply graded by seniority rather than by specific job function and tied to company profitability through bonus payments.

In short, they became members of the Toyota community, with a full set of rights, including the guarantee of lifetime employment and access to Toyota facilities (housing, recreation, and so forth), that went far beyond what most unions had been able to negotiate for mass-production employees in the West. In return, the company expected that most employees would remain with Toyota for their working lives.

This was a reasonable expectation because other Japanese companies adopted seniority-based wages at the same time, and workers would suffer a large loss of earnings if they started over at the bottom of the seniority ladder with another company. The wage progression was quite steep. A forty-year-old worker doing a given job received much higher pay than a twenty-five-year-old doing the same job. If the forty-year-old quit and went to work for another employer, he would start with a zero seniority wage that was below that of even the twenty-five-year-old.

The employees also agreed to be flexible in work assignments and active in promoting the interests of the company by initiating improvements rather than merely responding to problems. In effect, the company officials said: "If we are going to take you on for life, you have to do your part by doing the jobs that need doing," a bargain to which the unions agreed.

Back at the factory, Taiichi Ohno realized the implications of this historic settlement: The work force was now as much a short-term fixed cost as the company's machinery, and, in the long term, the workers were an even more significant fixed cost. After all, old machinery could be depreciated and scrapped, but Toyota needed to get the most out of its human resources over a forty-year period—that is, from the time new workers entered the company, which in Japan is generally between the ages of eighteen and twenty-two, until they reached retirement at age sixty. So it made sense to continuously enhance the workers' skills and to gain the benefit of their knowledge and experience as well as their brawn.

## LEAN PRODUCTION: FINAL ASSEMBLY PLANT

Ohno's rethinking of final assembly shows just how this new approach to human resources paid enormous dividends for Toyota. Remember that Ford's system assumed that assembly-line workers would perform one or two simple tasks, repetitively and, Ford hoped, without complaint. The foreman did not perform assembly tasks himself but instead ensured that the line workers followed orders. These orders or instructions were devised by the industrial engineer, who was also responsible for coming up with ways to improve the process.

Special repairmen repaired tools. Housekeepers periodically cleaned the work area. Special inspectors checked quality, and defective work, once discovered, was rectified in a rework area after the end of the line. A final category of worker, the utility man, completed the division of labor. Since even high wages were unable to prevent double-digit absenteeism in most mass-production assembly plants, companies needed a large group of utility workers on hand to fill in for those employees who didn't show up each morning.

Managers at headquarters generally graded factory management on two criteria—yield and quality. *Yield* was the number of cars actually produced in relation to the scheduled number, and *quality* was out-the-door quality, after vehicles had defective parts repaired. Factory managers knew that falling below the assigned production target spelled big trouble, and that mistakes could, if necessary, be fixed in the rework area, after the end of the line but before the cars reached the quality checker from headquarters stationed at the shipping dock. Therefore, it was crucial not to stop the line unless absolutely necessary. Letting cars go on down the line with a misaligned part was perfectly okay, because this type of defect could be rectified in the rework area, but minutes and cars lost to a line stoppage could only be made up with expensive overtime at the end of the shift. Thus was born the "move the metal" mentality of the mass-production auto industry.

Ohno, who visited Detroit repeatedly just after the war, thought this whole system was rife with *muda*, the Japanese term for waste that encompasses wasted effort, materials, and time. He reasoned that none of the specialists beyond the assembly worker was actually adding any value to the car. What's more, Ohno thought that assembly workers could probably do most of the functions of the specialists and do them much better because of their direct acquaintance with conditions on the line. (Indeed, he had just confirmed this observation in the press shop.) Yet, the role of the assembly worker had the lowest status in the factory. In some Western plants, management actually told assembly workers that they were needed only because automation could not yet replace them.

Back at Toyota City, Ohno began to experiment. The first step was to group workers into teams with a team leader rather than a foreman. The teams were given a set of assembly steps, their piece of the line, and told to work together on how best to perform the necessary operations. The team leader would do assembly tasks as well as coordinate the team, and, in particular, would fill in for any absent worker—concepts unheard of in mass-production plants.

Ohno next gave the team the job of housekeeping, minor tool repair, and quality-checking. Finally, as the last step, after the teams were running smoothly, he set time aside periodically for the team to suggest ways collectively to improve the process. (In the West, this collective suggestion process would come to be called "quality circles.") This continuous, incremental improvement process, *kaizen* in Japanese, took place in collaboration with the industrial engineers, who still existed but in much smaller numbers.

When it came to "rework," Ohno's thinking was truly inspired. He reasoned that the mass-production practice of passing on errors to keep the line running caused errors to multiply endlessly. Every worker could reasonably think that errors would be caught at the end of the line and that he was likely to be disciplined for any action that caused the line to stop. The initial error, whether a bad part or a good part improperly installed, was quickly compounded by assembly workers farther down the line. Once a defective part had become embedded in a complex vehicle, an enormous amount of rectification work might be

# THE RISE OF LEAN PRODUCTION

needed to fix it. And because the problem would not be discovered until the very end of the line, a large number of similarly defective vehicles would have been built before the problem was found.

So, in striking contrast to the mass-production plant, where stopping the line was the responsibility of the senior line manager, Ohno placed a cord above every work station and instructed workers to stop the whole assembly line immediately if a problem emerged that they couldn't fix. Then the whole team would come over to work on the problem.

Ohno then went much further. In mass-production plants, problems tended to be treated as random events. The idea was simply to repair each error and hope that it didn't recur. Ohno instead instituted a system of problem-solving called "the five why's." Production workers were taught to trace systematically every error back to its ultimate cause (by asking "why" as each layer of the problem was uncovered), then to devise a fix, so that it would never occur again.

Not surprisingly, as Ohno began to experiment with these ideas, his production line stopped all the time, and the workers easily became discouraged. However, as the work teams gained experience identifying and tracing problems to their ultimate cause, the number of errors began to drop dramatically. Today, in Toyota plants, where every worker can stop the line, yields approach 100 percent. That is, the line practically never stops! (In mass-production plants by contrast, where no one but the line manager can stop the line, the line still stops constantly. This is not to rectify mistakes—these are fixed at the end—but to deal with material supply and coordination problems. The consequence is that 90-percent yield is often taken as a sign of good management.)

Even more striking was what happened at the end of the line. As Ohno's system hit its stride, the amount of rework needed before shipment fell continually. Not only that, the quality of the shipped cars steadily improved. This was for the simple reason that quality inspection, no matter how diligent, simply cannot detect all the defects that can be assembled into today's complex vehicles.

Today, Toyota assembly plants have practically no rework areas and perform almost no rework. By contrast, as we will show, a number of current-day mass-production plants devote 20 percent of plant area and 25 percent of their total hours of effort to fixing mistakes. Perhaps the greatest testament to Ohno's ideas lies in the quality of the cars actually delivered to the consumer. American buyers report that Toyota's vehicles have among the lowest number of defects of any in the world, comparable to the very best of the German luxury car producers, who devote many hours of assembly-plant effort to rectification.

## LEAN PRODUCTION: THE SUPPLY CHAIN

Assembling the major components into a complete vehicle, the task of the final assembly plant, accounts for only 15 percent or so of the total manufacturing process. The bulk of the process involves engineering and fabricating more than 10,000 discrete parts and assembling these into perhaps 100 major components—engines, transmissions, steering gears, suspensions, and so forth.

Coordinating this process so that everything comes together at the right time with high quality and low cost has been a continuing challenge to the final assembler firms in the auto industry. Under mass production, as we noted earlier, the initial intention was to integrate the entire production system into one huge, bureaucratic command structure with orders coming down from the top. However, even Alfred Sloan's managerial innovations were unequal to this task.

The world's mass-production assemblers ended up adopting widely varying degrees of formal integration, ranging from about 25 percent in-house production at small specialist firms, such as Porsche and Saab, to about 70 percent at General Motors. Ford, the early leader in vertical integration, which actually did approach 100 percent at the Rouge, deintegrated after World War II to about 50 percent.

However, the make-or-buy decisions that occasioned so much debate in mass-production firms struck Ohno and others at Toyota as largely irrelevant, as they began to consider obtaining components for cars and trucks. The real question was how the assembler and the suppliers could work smoothly together to reduce costs and improve quality, whatever formal, legal relationship they might have.

And here the mass-production approach—whether to make or buy—seemed broadly unsatisfactory. At Ford and GM, the central engineering staffs designed most of the 10,000-plus parts in a vehicle and the component systems they comprised. The firms then gave the drawings to their suppliers, whether formally part of the assembler firm or independent businesses, and asked them for bids on a given number of parts of given quality (usually expressed as a maximum number of defective parts per 1,000) delivered at a given time. Among all the outside firms and internal divisions that were asked to bid, the low bidder got the business.[10]

For certain categories of parts, typically those shared by many vehicles (tires, batteries, alternators) or involving some specialized technology that the assembler firm didn't have (engine computers, for example), independent supplier firms competed to supply the parts, usually by modifying existing standard designs to meet the specifications of a particular vehicle. Again, success depended upon price, quality, and delivery reliability, and the car makers often switched business between firms on relatively short notice.

In both cases, corporate managers and small-business owners alike understood that it was every firm for itself when sales declined in the cyclical auto industry. Everyone thought of their business relationships as characteristically short-term.

As the growing Toyota firm considered this approach to components supply, Ohno and others saw many problems. Supplier organizations, working to blueprint, had little opportunity or incentive to suggest improvements in the production design based on their own manufacturing experience. Like employees in the mass-production assembly plant, they were told in effect to keep their heads down and continue working. Alternatively, suppliers offering standardized designs of their own, modified to specific vehicles, had no practical way of optimizing these parts, because they were given practically no information about the rest of the vehicle. Assemblers treated this information as proprietary.

And there were other difficulties. Organizing suppliers in vertical chains and playing them against each other in search of the lowest short-term cost blocked the flow of information horizontally between suppliers, particularly on advances in manufacturing techniques. The assembler might ensure that suppliers had low profit margins, but not that they steadily decreased the cost of production through improved organization and process innovations.

The same was true of quality. Because the assembler really knew very little about its suppliers' manufacturing techniques—whether the supplier in question was inside the assembler firm or independent—it was hard to improve quality except by establishing a maximum acceptable level of defects. As long as most firms in the industry produced to about the same level of quality, it was difficult to raise that level.

Finally, there was the problem of coordinating the flow of parts within the supply system on a day-to-day basis. The inflexibility of tools in supplier plants (analogous to the inflexibility of the stamping presses in the assembler plants) and the erratic nature of orders from assemblers responding to shifting market demand caused suppliers to build large volumes of one type of part before changing over machinery to the next and to maintain large stocks of finished parts in a warehouse so that the assembler would never have cause to complain (or worse, to cancel a contract) because of a delay in delivery. The result was high inventory costs and the routine production of thousands of parts that were later found to be defective when installed at the assembly plant.

To counteract these problems and to respond to a surge in demand in the 1950s, Toyota began to establish a new, lean-production approach to components supply. The first step was to organize suppliers into functional tiers, whatever the legal, formal relation of the supplier to the assembler. Different responsibilities were assigned to firms in each tier. First-tier suppliers were responsible for working as an integral part of the product-development team in developing a new product. Toyota told them to develop, for example, a steering, braking, or electrical system that would work in harmony with the other systems.

First, they were given a performance specification. For example, they were told to design a set of brakes that could stop a 2,200-pound car from 60 miles per hour in 200 feet ten times in succession without fading. The brakes should fit into a space 6" × 8" × 10" at the end of each axle and be delivered to the assembly plant for $40 a set. The suppliers were then told to deliver a prototype for testing. If the prototype worked, they got a production order. Toyota did not specify what the brakes were made of or how they were to work. These were engineering decisions for the supplier to make.

Toyota encouraged its first-tier suppliers to talk among themselves about ways to improve the design process. Because each supplier, for the most part, specialized in one type of component and did not compete in that respect with other suppliers in the group, sharing this information was comfortable and mutually beneficial.

Then, each first-tier supplier formed a second tier of suppliers under itself. Companies in the second tier were assigned the job of fabricating individual parts. These suppliers were manufacturing specialists, usually without much expertise in product engineering but with strong backgrounds in process engineering and plant operations.

For example, a first-tier supplier might be responsible for manufacturing alternators. Each alternator has around 100 parts, and the first-tier supplier would obtain all of these parts from second-tier suppliers.

Because second-tier suppliers were all specialists in manufacturing processes and not competitors in a specific type of component, it was easy to group them into supplier associations so that they, too, could exchange information on advances in manufacturing techniques.

Toyota did not wish to vertically integrate its suppliers into a single, large bureaucracy. Neither did it wish to deintegrate them into completely independent companies with only a marketplace relationship. Instead, Toyota spun its in-house supply operations off into quasi-independent first-tier supplier companies in which Toyota retained a fraction of the equity and developed similar relationships with other suppliers who had been completely independent. As the process proceeded, Toyota's first-tier suppliers acquired much of the rest of the equity in each other.

Toyota, for instance, today holds 22 percent of Nippondenso, which makes electrical components and engine computers; 14 percent of Toyoda Gosei, which makes seats and wiring systems; 12 percent of Aishin Seiki, which makes metal engine parts; and 19 percent of Koito, which makes trim items, upholstery, and plastics. These firms, in turn, have substantial cross-holdings in each other. In addition, Toyota often acts as banker for its supplier group, providing loans to finance the process machinery required for a new product.

Finally, Toyota shared personnel with its supplier-group firms in two ways. It would lend them personnel to deal with workload surges, and it would transfer senior managers not in line for top positions at Toyota to senior positions in supplier firms.

Consequently, the Toyota suppliers were independent companies, with completely separate books. They were real profit centers, rather than the sham profit centers of many vertically integrated mass-production firms. Moreover, Toyota encouraged them to perform considerable work for other assemblers and for firms in other industries, because outside business almost always generated higher profit margins. (Nippondenso, for example, a $7-billion company, is the largest manufacturer in the world of electrical and electronic systems and engine computers. As we mentioned, Toyota holds 22 percent of its equity, and Nippondenso does 60 percent of its business with Toyota. Probably another 30 percent of Nippondenso's equity is held in the Toyota supplier group of companies, and 6 percent is held by Robert Bosch, the giant German components firm. The rest trades publicly.)

At the same time, these suppliers are intimately involved in Toyota's product development, have interlocking equity with Toyota and Toyota group members, rely on Toyota for outside financing, and accept Toyota people into their personnel systems. In a very real sense, they share their destinies with Toyota.

Finally, Ohno developed a new way to coordinate the flow of parts within the supply system on a day-to-day basis, the famous just-in-time system, called *kanban* at Toyota. Ohno's idea was simply to convert a vast group of suppliers and parts plants into one large machine, like Henry Ford's Highland Park plant, by dictating that parts would only be produced at each previous step to supply the immediate demand of the next step. The mechanism was the containers carrying parts to the next step. As each container was used up, it was sent back to the previous step, and this became the automatic signal to make more parts.[11]

This simple idea was enormously difficult to implement in practice because it eliminated practically all inventories and meant that when one small part of the vast production system failed, the whole system came to a stop. In Ohno's view, this was precisely the power of his idea—it removed all safety nets and focused every member of the vast production process on anticipating problems before they became serious enough to stop everything.

It took Eiji Toyoda and Ohno more than twenty years of relentless effort to fully implement this full set of ideas—including just-in-time—within the Toyota supply chain. In the end they succeeded, with extraordinary consequences for productivity, product quality, and responsiveness to changing market demand. As we'll see in chapters 4 and 5, the lean supply chain became a major strength of the lean production system.

## LEAN PRODUCTION: PRODUCT DEVELOPMENT AND ENGINEERING

Wherever it occurs—at central engineering headquarters or in the supplier organizations—the process of engineering a manufactured object as complex as today's motor vehicle demands enormous effort from large numbers of people with a broad range of skills. Therefore, it's easy to make mistakes in organizing the process so that the whole of the results achieved is mysteriously less than the sum of the parts.

Mass-production companies try to solve the complexity problem by finely dividing labor among many engineers with very specific specialties. Professor Kim Clark of the Harvard Business School reports, for instance, finding an engineer in a mass-production auto company who had spent his whole career designing auto door locks. He was not an expert on how to *make* door locks, however; that was the job of the door-lock manufacturing engineer. The door-lock design engineer simply knew how they should look and work if made correctly.

The weaknesses of this system of divided labor were easy to see and mass-production companies over the years tried to devise coordination mechanisms. Even in the mid-1980s, though, the best solution they had found was the product-development team with a weak leader (really just a coordinator), whose members still reported to the senior executive of their individual technical specialties. Significantly, career paths in most Western firms still followed a constricted progression through their technical department: junior piston engineer to senior piston engineer, and then junior drive-train engineer to senior drive-train engineer, and so forth. One might someday hope to reach the position of chief product engineer, the level at which disagreements among product engineers, manufacturing-process engineers, and industrial engineers in the plants were worked out.

Ohno and Toyoda, by contrast, decided early on that product engineering inherently encompassed both process and industrial engineering. Thus, they formed teams with strong leaders that contained all the relevant expertise. Career paths were structured so that rewards went to strong team players rather than to those displaying genius in a single area of product, process, or industrial engineering, but without regard to their function as a team.

As we'll see in Chapter 5, the consequence of this approach to lean engineering was a dramatic leap in productivity, product quality, and responsiveness to changing consumer demand.

## LEAN PRODUCTION AND CHANGING CONSUMER DEMAND

The new Toyota production system was especially well suited to capitalize upon the changing demands that consumers were placing on their cars and upon changing vehicle technology. By the 1960s, cars and light trucks were increasingly a part of daily life in developed countries. Almost everyone, even people with no inherent interest in cars, depended on them to get through the day.

Simultaneously, vehicles were acquiring features that made them quite impossible for the average user to repair. The putty knife and wrench that could fix almost anything that could go wrong with a Model T were of little use by the 1980s for a broken engine-management computer or antiskid braking system.

Also, as households began to acquire more than one vehicle, people no longer wanted just the standard-size car or truck. The market began to fragment into many product segments.

For the Toyota production system, these developments were all blessings: Consumers began to report that the most important feature of their car or truck was reliability. It had to start every morning and could never leave the user stranded. Vehicle malfunctions were no longer a challenge for the home tinkerer, but were inexplicable nightmares, even for owners with considerable mechanical skills. Because the Toyota system could deliver superior reliability, soon Toyota found that it no longer had to match exactly the price of competing mass-production products.

Furthermore, Toyota's flexible production system and its ability to reduce production-engineering costs let the company supply the product variety that buyers wanted with little cost penalty. In 1990, Toyota is offering consumers around the world as many products as General Motors—even though Toyota is still half GM's size. To change production and model specifications in mass-production firms takes many years and costs a fortune. By contrast, a preeminent lean producer, such as Toyota, needs half the time and effort required by a mass-producer such as GM to design a new car. So Toyota can offer twice as many vehicles with the same development budget.

Ironically, most Western companies concluded that the Japanese succeeded because they produced standardized products in ultra-high volume. As recently as 1987 a manufacturing manager in Detroit confided in an interview with members of our project that he had found the secret of Japanese success: "They are making identical tin cans; if I did that I could have high quality and low cost, too." This illusion stems from the fact that the Japanese companies initially minimized distribution costs by focusing on one or two product categories in each export market.

However, the total product portfolio of the Japanese firms has always been broader, and they have steadily increased their product range in every world market. Today they offer almost as many models as all of the Western firms combined, as we will see in Chapter 5.[12] In addition, their product variety continues to grow rapidly even as Western firms stand still on average and actually reduce the number of different models made in each of their factories. Ford and GM, for example, have been "focusing" their assembly plants toward the goal of one basic product in each plant. By contrast, the Japanese transplants in North America all build two or three products.

Because product lives now average just four years, the average production volume of a Japanese car over the period of its production is now one-quarter that of the Western mass-market producers, and the gap is widening. That is, the Japanese currently make, on average, 125,000 copies of each of their car models each year. The seven Western high-volume companies make nearly twice as many. However, the Japanese keep models in production four years on average, while the Western companies keep them in production almost ten years. This means that during the life of a model the Japanese make 500,000 copies (125,000 times 4), while the Western companies make 2 million (200,000 times 10), a four-to-one difference.

Even more striking, Japanese producers such as Toyota are already producing at only two-thirds of the life-of-the-model production volume of European specialist firms, such as Mercedes and BMW. Indeed, with the arrival of a host of new Japanese niche cars, such as the Honda NS-X, the Japanese may be able to do what mass-production firms never could: attack the surviving craft-based niche producers, such as Aston Martin and Ferrari, to bring the whole world into the age of lean production.

## LEAN PRODUCTION: DEALING WITH THE CUSTOMER

All of the variety available from lean production would be for naught if the lean producer could not build what the customer wanted. Thus from an early date Eiji Toyoda and his marketing expert, Shotaro Kamiya, began to think about the link between the production system and the consumer.

For Henry Ford this link had been very simple: Because there was no product variety and because most repairs could be handled by the owner, the job of the dealer was simply to have enough cars and spare parts in stock to supply expected demand. In addition, because demand in the American car market fluctuated wildly from the earliest days of the industry, the assembler tended to use the dealer as a shock absorber to cushion the factory from the need to increase and reduce production continually. The result, fully in place by the 1920s, was a system of small, financially independent dealers who maintained a vast inventory of cars and trucks waiting for buyers.

Relations between the factory and the dealer were distant and usually strained as the factory tried to force cars on dealers to smooth production. Relations between the dealer and the customer were equally strained because dealers continually adjusted prices—made deals—to adjust demand with supply while maximizing profits. As anyone who has bought a car in North America or Europe knows, this has been a system marked by a lack of long-term commitment on either side, which maximizes feelings of mistrust. In order to maximize bargaining position, everyone holds back information—the dealer about the product, the consumer about his or her true desires—and everyone loses in the long term.

Kamiya had learned this system by working in General Motor's Japanese distribution system in the 1930s, but it seemed broadly unsatisfactory. Therefore, after the war he and Toyoda began to think of new ways to distribute cars.[13] Their solution, worked out gradually over time, was to build a sales network very similar to the Toyota supplier group, a system that had a very different relation with the customer.

Specifically, the Toyota Motor Sales Company[14] built up a network of distributors, some wholly owned and some in which Toyota held a small equity stake, who had a "shared destiny" with Toyota. These dealers developed a new set of techniques that Toyota came to call "aggressive selling." The basic idea was to develop a long-term, indeed a life-long, relation between the assembler, the dealer, and the buyer by building the dealer into the production system and the buyer into the product development process.

The dealer became part of the production system as Toyota gradually stopped building cars in advance for unknown buyers and converted to a build-to-order system in which the dealer was the first step in the *kanban* system, sending orders for presold cars to the factory for delivery to specific customers in two to three weeks. To make this workable, however, the dealer had to work closely with the factory to sequence orders in a way the factory could accommodate. While Ohno's production system was remarkably adept at building products to specific order, it could not deal with large surges or troughs in total demand or abrupt shifts in demand between products that could not be built with the same tools—for example, between the largest and smallest cars in the product range or between cars and trucks.

Sequencing orders was possible in turn because Toyota's sales staff did not wait in the showroom for orders. Instead they went directly to customers by making house calls. When demand began to droop they worked more hours, and when demand shifted they concentrated on households they knew were likely to want the type of car the factory could build.

The latter was possible because of a second feature of aggressive selling—a massive data base on households and their buying preferences that Toyota gradually built up on every household ever showing interest in a Toyota product. With this information in hand, Toyota sales staff could target their efforts to the most likely buyers.

The system also could incorporate the buyer into the product development process and in a very direct way. Toyota focused relentlessly on repeat buyers—critical in a country where government vehicle inspections, the famous *shoken,* resulted in practically every car being scrapped after six years. Toyota was determined never to lose a former buyer and could minimize the chance of this happening by using the data in its consumer data base to predict what Toyota buyers would want next as their incomes, family size, driving patterns, and tastes changed. Unlike mass-producers who conduct product evaluation "clinics" and other survey research on randomly selected buyers—buyers presumed to have little "brand loyalty"—Toyota went directly to its existing customers in planning new products. Established customers were treated as members of the "Toyota family," and brand loyalty became a salient feature of Toyota's lean production system.

## THE FUTURE OF LEAN PRODUCTION

Toyota had fully worked out the principles of lean production by the early 1960s. The other Japanese auto firms adopted most of them as well, although this took many years. For example, Mazda did not fully embrace Ohno's ideas for running factories and the supplier system until it encountered a crisis in 1973, when export demand for its fuel-hungry Wankel-engined cars collapsed. The first step of the Sumitomo group in offering help to Mazda was to insist that the company's Hiroshima production complex rapidly remake itself in the image of Toyota City at Nagoya.

What's more, not all firms became equally adept at operating the system. (One our most important objectives in this volume is to educate the public to the fact that some Japanese firms are leaner than others and that several of the old-fashioned mass-production firms in the West are rapidly becoming lean as well.) Nevertheless, by the 1960s the Japanese firms on average had gained an enormous advantage over mass-producers elsewhere and were able for a period of twenty years to boost their share of world motor vehicle production steadily by exporting from their highly focused production complexes in Japan, as shown in Figure 3.1.

This path of export-led development came to an abrupt halt after 1979, as the world economy slumped, trade imbalances with North America and Europe reached unmanageable proportions, and trade barriers were erected. In the 1980s the world was at the same point in the diffusion of lean production that it was with mass production in the 1920s: The leading practitioners of the new method are now of necessity attempting to increase their share of the world market through direct investment in North America and Europe (as shown in the checked area in Figure 3.1) rather than through ever growing exports of finished units. Meanwhile, American, European, and even Korean firms—often accom-

## THE RISE OF LEAN PRODUCTION

**FIGURE 3.1**

**Japanese Share of World Motor Vehicle Production, 1955–1989**

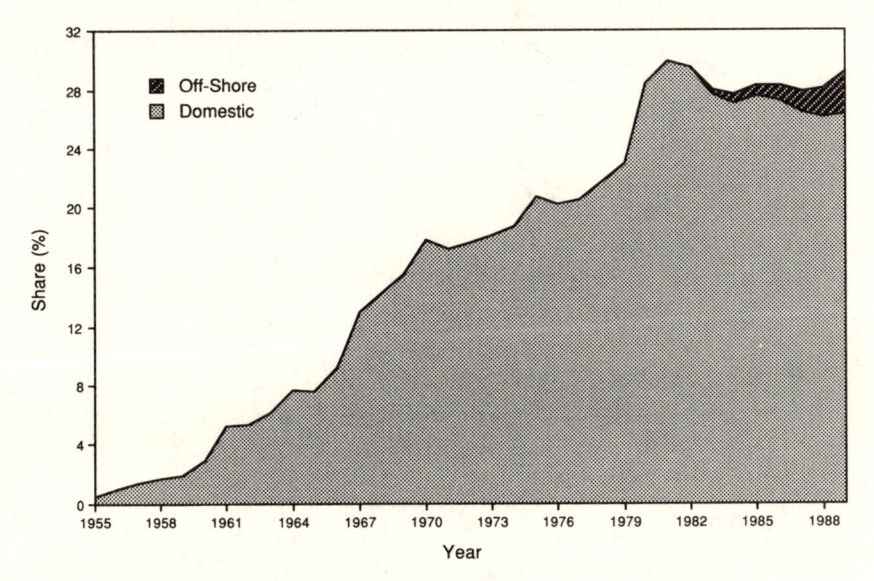

*Note:* Includes both domestic and off-shore production.
*Source: Automotive News Market Data Book*

plished masters of now obsolete mass production—are trying to match or exceed the performance of their lean challengers.

This process is enormously exciting. It also produces enormous tensions. There will be real losers (including some of the smaller and less accomplished Japanese firms) as well as winners, and the public everywhere tends all too readily to interpret the contest in simple nationalistic terms—"us" versus "them," "our" country versus "theirs."

We will return to the problem of diffusing lean production in the final chapters of this book because we believe it is one of the most important issues facing the world economy in the 1990s. However, we first need to gain a deeper understanding of the elements of lean production.

# THE ELEMENTS OF LEAN PRODUCTION

The general public has a simple and vivid mental image of auto production—the assembly plant where all the parts come together to create the finished car or truck. While this final manufacturing step is important, it represents only about 15 percent of the human effort involved in making a car. To properly understand lean production, we must look at every step in the process, beginning with product design and engineering, then going far beyond the factory to the customer who relies on the automobile for daily living. In addition, it is critical to understand the mechanism of coordination necessary to bring all these steps into harmony and on a global scale, a mechanism we call *the lean enterprise*.

In the chapters just ahead, we proceed through each of the steps of lean production. We begin with the part of the system everyone thinks they understand—the factory as represented by the assembly plant—to show systematically how very different lean production is from the ideas of Henry Ford. We proceed to product development and engineering, then into the supply system, where the bulk of manufacturing occurs. Next we look at the system of selling cars and trucks—the end of the production process in the world of mass production, but the beginning of the process in lean production. Finally, we examine the type of global lean enterprise needed to make the whole system work, the one aspect of lean production that is still not fully developed.

# RUNNING THE FACTORY

## 4

The automotive assembly plant dominates its landscape, wherever in the world it's located. From a distance, it is a vast windowless mass surrounded by acres of storage areas and railway yards. The complex shape of the building and the lack of a facade often make it hard to know just where to enter. Once inside, the scene is initially bewildering.

Thousands of workers in one vast building tend to streams of vehicles moving across the floor, while a complex network of conveyors and belts in the lofty ceiling carries parts to and fro. The scene is dense, hectic, noisy. On first exposure, it's like finding oneself inside a Swiss watch—fascinating but incomprehensible and a little frightening, as well.

In 1986, at the outset of the IMVP, we set out to contrast lean production with mass production by carefully surveying as many of the world's motor-vehicle assembly plants as possible. In the end, we visited and systematically gathered information on more than ninety plants in seventeen countries, or about half of the assembly capacity of the entire world. Ours would prove one of

---

This chapter is based on the IMVP World Assembly Plant Survey. The survey was initiated by John Krafcik, who was later joined by John Paul MacDuffie. Haruo Shimada also assisted.

the most comprehensive international surveys ever undertaken in the automobile or any other industry.

Why did we choose the assembly plant for study? Why not the engine plant, say, or the brake plant or the alternator factory? And why so many plants in so many countries? Surely, the best lean-production plant in Japan and the worst mass-production plant in North America or Europe would have sufficiently demonstrated the differences between lean and mass production.

Three factors convinced us that the assembly plant was the most useful activity in the motor-vehicle production system to study.

First, a large part of the work in the auto industry involves assembly. This is so simply because of the large number of parts in a car. Much of this assembly occurs in components plants. For example, an alternator plant will gather from suppliers or fabricate the 100 or so discrete parts in an alternator, then assemble them into a complete unit. However, it's hard to understand assembly in such a plant, because the final activity usually makes up only a small part of the total. In the final assembly plant, by contrast, the sole activity is assembly—welding and screwing several thousand simple parts and complex components into a finished vehicle.

Second, assembly plants all over the world do almost exactly the same things, because practically all of today's cars and light trucks are built with very similar fabrication techniques. In almost every assembly plant, about 300 stamped steel panels are spot-welded into a complete body. Then the body is dipped and sprayed to protect it from corrosion. Next, it is painted. Finally, thousands of mechanical parts, electrical items, and bits of upholstery are installed inside the painted body to produce the complete automobile. Because these tasks are so uniform, we can meaningfully compare a plant in Japan with one in Canada, another in West Germany, and still another in China, even though they are making cars that look very different as they emerge from the factory.

Finally, we chose the assembly plant for study, because Japanese efforts to spread lean production by building plants in North America and Europe initially involved assembly plants. When we began our survey in 1986, three Japanese-managed assembly plants were already in operation in the United States and one was ready to open in England.

By contrast, Japanese plants for engines, brakes, alternators, and other components, though publicly announced for North America and Europe, were still in the planning stage. We knew from experience that it's pointless to examine a company's blueprints for a new plant or to look at a plant just as it starts production. To see the full difference between lean and mass production at the plant level, we had to compare plants operating at full volume.

What about the second question we are often asked: "Why study so many plants in so many countries?" The answer is simple. Lean production is now spreading from Japan to practically every nation. Directly in its path are the giant mass-production plants of the previous industrial era.

In every country and every company—including, we might add, in the less accomplished companies in Japan—we have found an intense, even desperate, desire to know the answer to two simple questions: "Where do we stand?" and "What must we do to match the new competitive level required by lean production?" Now we know the answers.

## CLASSIC MASS PRODUCTION: GM FRAMINGHAM

We began our survey in 1986 at General Motors' Framingham, Massachusetts, assembly plant, just a few miles south of our home base in Boston. We chose Framingham, not because it was nearby, but because we strongly suspected it embodied all the elements of classic mass production.

Our first interview with the plant's senior managers was not promising. They had just returned from a tour of the Toyota-GM joint-venture plant (NUMMI) where John Krafcik, our assembly plant survey leader, formerly worked. One reported that secret repair areas and secret inventories had to exist behind the NUMMI plant, because he hadn't seen enough of either for a "real" plant. Another manager wondered what all the fuss was about. "They build cars just like we do." A third warned that "all that NUMMI talk [about lean production] is not welcome around here."

Despite this cold beginning, we found the plant management enormously helpful. All over the world, as we have since discovered again and again, managers and workers badly want to learn where they stand and how to improve. Their fear of just how bad things might be is in fact what often creates initial hostility.

On the plant floor, we found about what we had expected: a classic mass-production environment with its many dysfunctions. We began by looking down the aisles next to the assembly line. They were crammed with what we term *indirect workers*—workers on their way to relieve a fellow employee, machine repairers en route to troubleshoot a problem, housekeepers, inventory runners. None of these workers actually add value, and companies can find other ways to get their jobs done.

Next, we looked at the line itself. Next to each work station were piles—in some cases weeks' worth—of inventory. Littered about were discarded boxes and other temporary wrapping material. On the line itself the work was unevenly distributed, with some workers running madly to keep up and others finding time to smoke or even read a newspaper. In addition, at a number of points the workers seemed to be struggling to attach poorly fitting parts to the Oldsmobile Ciera models they were building. The parts that wouldn't fit at all were unceremoniously chucked in trash cans.

At the end of the line we found what is perhaps the best evidence of old-fashioned mass production: an enormous work area full of finished cars riddled with defects. All these cars needed further repair before shipment, a task that can prove enormously time-consuming and often fails to fix fully the problems now buried under layers of parts and upholstery.

On our way back through the plant to discuss our findings with the senior managers, we found two final signs of mass production: large buffers of finished bodies awaiting their trip through the paint booth and from the paint booth to the final assembly line, and massive stores of parts, many still in the railway cars in which they had been shipped from General Motors' components plants in the Detroit area.

Finally, a word on the work force. *Dispirited* is the only label that would fit. Framingham workers had been laid off a half-dozen times since the beginning of the American industry's crisis in 1979, and they seemed to have little hope that the plant could long hold out against the lean-production facilities locating in the American Midwest.

# RUNNING THE FACTORY

## CLASSIC LEAN PRODUCTION: TOYOTA TAKAOKA

Our next stop was the Toyota assembly plant at Takaoka in Toyota City. Like Framingham (built in 1948), this is a middle-aged facility (from 1966). It had a much larger number of welding and painting robots in 1986 but was hardly a high-tech facility of the sort General Motors was then building for its new GM-10 models, in which computer-guided carriers replaced the final assembly line.

The differences between Takaoka and Framingham are striking to anyone who understands the logic of lean production. For a start, hardly anyone was in the aisles. The armies of indirect workers so visible at GM were missing, and practically every worker in sight was actually adding value to the car. This fact was even more apparent because Takaoka's aisles are so narrow.

Toyota's philosophy about the amount of plant space needed for a given production volume is just the opposite of GM's at Framingham: Toyota believes in having as little space as possible so that face-to-face communication among workers is easier, and there is no room to store inventories. GM, by contrast, has believed that extra space is necessary to work on vehicles needing repairs and to store the large inventories needed to ensure smooth production.

The final assembly line revealed further differences. Less than an hour's worth of inventory was next to each worker at Takaoka. The parts went on more smoothly and the work tasks were better balanced, so that every worker worked at about the same pace. When a worker found a defective part, he—there are no women working in Toyota plants in Japan—carefully tagged it and sent it to the quality-control area in order to obtain a replacement part. Once in quality control, employees subjected the part to what Toyota calls "the five why's" in which, as we explained in Chapter 2, the reason for the defect is traced back to its ultimate cause so that it will not recur.

As we noted, each worker along the line can pull a cord just above the work station to stop the line if any problem is found; at GM only senior managers can stop the line for any reason other than safety—but it stops frequently due to problems with machinery or materials delivery. At Takaoka, every worker can stop the line but the line is almost never stopped, because problems are solved in advance and the same problem never occurs twice. Clearly, paying relentless attention to preventing defects has removed most of the reasons for the line to stop.

At the end of the line, the difference between lean and mass production was even more striking. At Takaoka, we observed almost no rework area at all. Almost every car was driven directly from the line to the boat or the trucks taking cars to the buyer.

On the way back through the plant, we observed yet other differences between this plant and Framingham. There were practically no buffers between the welding shop and paint booth and between paint and final assembly. And there were no parts warehouses at all. Instead parts were delivered directly to the line at hourly intervals from the supplier plants where they had just been made. (Indeed, our initial plant survey form asked how many days of inventory were in the plant. A Toyota manager politely asked whether there was an error in translation. Surely we meant *minutes* of inventory.)

A final and striking difference with Framingham was the morale of the work force. The work pace was clearly harder at Takaoka, and yet there was a sense of purposefulness, not simply of workers going through the motions with their minds elsewhere under the watchful eye of the foreman. No doubt this was in considerable part due to the fact that all of the Takaoka workers were lifetime employees of Toyota, with fully secure jobs in return for a full commitment to their work.[1]

## A BOX SCORE: MASS PRODUCTION VERSUS LEAN

When the team had surveyed both plants, we began to construct a simple box score to tell us how productive and accurate each plant was ("accurate" here means the number of assembly defects in cars as subsequently reported by buyers).[2] It was easy to calculate a gross productivity comparison, dividing the number of hours worked by all plant employees by the number of vehicles produced, as shown in the top line of Figure 4.1[3] However, we had to make sure that each plant was performing exactly the same tasks. Otherwise, we wouldn't be comparing apples with apples.

So we devised a list of standard activities for both plants—welding of all body panels, application of three coats of paint, installation of all parts, final inspection, and rework—and noted

**FIGURE 4.1**

**General Motors Framingham Assembly Plant Versus Toyota Takaoka Assembly Plant, 1986**

|  | GM Framingham | Toyota Takaoka |
|---|---|---|
| Gross Assembly Hours per Car | 40.7 | 18.0 |
| Adjusted Assembly Hours per Car | 31 | 16 |
| Assembly Defects per 100 Cars | 130 | 45 |
| Assembly Space per Car | 8.1 | 4.8 |
| Inventories of Parts (average) | 2 weeks | 2 hours |

*Note:* Gross assembly hours per car are calculated by dividing total hours of effort in the plant by the total number of cars produced.

"Adjusted assembly hours per car" incorporates the adjustments in standard activities and product attributes described in the text.

Defects per car were estimated from the J. D. Power Initial Quality Survey for 1987.

Assembly space per car is square feet per vehicle per year, corrected for vehicle size.

Inventories are a rough average for major parts.

*Source:* IMVP World Assembly Plant Survey

---

any task one plant was doing that the other wasn't. For example, Framingham did only half its own welding and obtained many prewelded assemblies from outside contractors. We made an adjustment to reflect this fact.

We also knew it would make little sense to compare plants assembling vehicles of grossly different sizes and with differing amounts of optional equipment, so we adjusted the amount of effort in each plant as if a standard vehicle of a specified size and option content were being assembled.[4]

When our task was completed, an extraordinary finding emerged, as shown in Figure 4.1. Takaoka was almost twice as productive and three times as accurate as Framingham in performing the same set of standard activities on our standard car. In terms of manufacturing space, it was 40 percent more efficient, and its inventories were a tiny fraction of those at Framingham.

If you remember Figure 2.1 from Chapter 2, you might wonder whether this leap in performance from classic mass production, as practiced by GM, to classic lean production, as performed by Toyota, really deserves the term *revolution*. After all, Ford managed to reduce direct assembly effort by a factor of nine at Highland Park.

In fact, Takaoka is in some ways an even more impressive achievement than Ford's at Highland Park, because it represents an advance on so many dimensions. Not only is effort cut in half and defects reduced by a factor of three, Takaoka also slashes inventories and manufacturing space. (That is, it is both capital- and labor-saving compared with Framingham-style mass production.) What's more, Takaoka is able to change over in a few days from one type of vehicle to the next generation of product while Highland Park, with its vast array of dedicated tools, was closed for months in 1927 when Ford switched from the Model T to the new Model A. Mass-production plants continue to close for months while switching to new products.

## DIFFUSING LEAN PRODUCTION

Revolutions in manufacture are useful only if they are available to everyone. We were therefore vitally interested in learning if the new transplant facilities being opened in North America and Europe could actually institute lean production in a different environment.

We knew one of the Japanese transplants in North America very well, of course, because of IMVP research affiliate John Krafcik's tenure there. The New United Motor Manufacturing Inc. (NUMMI) plant in Fremont, California, is a joint venture between the classic mass-producer, GM, and the classic lean producer, Toyota.

NUMMI uses an old General Motors plant built in the 1960s to assemble GM cars and pickup trucks for the U.S. West Coast. As GM's market share along the Pacific Coast slipped steadily, the plant had less and less work. It finally closed its doors for good in 1982. By 1984 GM had decided that it needed to learn about lean production from the master. So it convinced Toyota to provide the management for a reopened plant, which would produce small Toyota-designed passenger cars for the U.S. market.

NUMMI was to make no compromises on lean production. The senior managers were all from Toyota and quickly imple-

*RUNNING THE FACTORY*

mented an exact copy of the Toyota Production System. A key action toward this end was the construction of a new stamping plant adjacent to the body-welding area, so that body panels could be stamped in small lots just as they were needed. By contrast, the old Fremont plant had depended on panels supplied by rail from GM's centralized stamping plants in the Midwest. There they were stamped out by the million on dedicated presses.

The United Automobile Workers Union also cooperated to make lean production possible. Eighty percent of the NUMMI work force consisted of workers formerly employed by GM at Fremont. However, in place of the usual union contract with thousands of pages of fine print defining narrow job categories and other job-control issues, the NUMMI contract provided for only two categories of workers—assemblers and technicians. The union agreed as well that all its members should work in small teams to get the job done with the least effort and highest quality.

By the fall of 1986, NUMMI was running full blast. And we were ready to compare it with Takaoka and Framingham, as shown in Figure 4.2.

We found that NUMMI matched Takaoka's quality and nearly matched its productivity. Space utilization was not as efficient because of the old GM plant's poor layout. Inventory was also considerably higher than at Takaoka, because almost all the parts were transported 5,000 miles across the Pacific rather than five or ten miles from neighboring supplier plants in Toyota City. (Even so, NUMMI was able to run with a two-day supply of parts, while Framingham needed two weeks' worth.)

**FIGURE 4.2**

**General Motors Framingham versus Toyota Takaoka versus NUMMI Fremont, 1987**

|  | GM Framingham | Toyota Takaoka | NUMMI Fremont |
|---|---|---|---|
| Assembly Hours per Car | 31 | 16 | 19 |
| Assembly Defects per 100 Cars | 135 | 45 | 45 |
| Assembly Space per Car | 8.1 | 4.8 | 7.0 |
| Inventories of Parts (average) | 2 weeks | 2 hours | 2 days |

*Source:* IMVP World Assembly Plant Survey

It was clear to us by the end of 1986 that Toyota had truly achieved a revolution in manufacturing, that old mass-production plants could not compete, and that the new best way—lean production—could be transplanted successfully to new environments, such as NUMMI. Given these findings, we were hardly surprised by subsequent events: Takaoka continues to improve, now with much additional automation. NUMMI is also getting continually better and a second line is being added to assemble Toyota pickup trucks. Framingham was closed forever in the summer of 1989.

## SURVEYING THE WORLD

Once we finished our initial survey, we were determined to press ahead on a survey of the entire world. We were motivated partly by the fact that the companies and governments sponsoring us wanted to know where they stood and partly by the knowledge that a survey of three plants could not answer a number of questions about what roles automation, manufacturability, product variety, and management practices play in the success of manufacturing.

However, we soon realized that we would have to conceal company and plant names when we reported our findings. Many companies were willing to grant us access to their plants only on the condition that we would not reveal plant names in our results. We have respected their wishes and in this book identify plants only when the company has agreed.

After four more years of research, we have found the following about productivity and quality (or accuracy) across the world, as summarized in figures 4.3 and 4.4.

These findings are not at all what we had expected. We had anticipated all of the Japanese firms in Japan to be roughly comparable in performance—that is, equally lean. Further, we had expected all of the American plants in North America and the American- and European-owned plants in Europe to perform at about the same level with little variation and to trail the average Japanese plant by about the same degree that Framingham trailed Takaoka in 1986. Finally, we expected the assembly plants in developing countries to be marked by low productivity and low quality. The reality is different.

# RUNNING THE FACTORY

**FIGURE 4.3**

**Assembly Plant Productivity, Volume Producers, 1989**

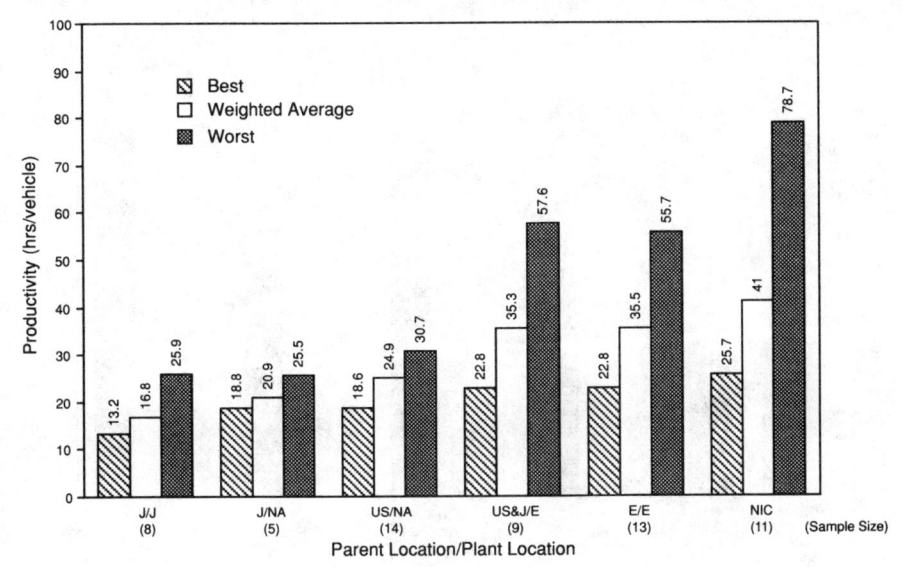

[Bar chart showing productivity (hrs/vehicle) by Parent Location/Plant Location, with Best, Weighted Average, and Worst categories]

| Category | Best | Weighted Average | Worst |
|---|---|---|---|
| J/J (8) | 13.2 | 16.8 | 25.9 |
| J/NA (5) | 18.8 | 20.9 | 25.5 |
| US/NA (14) | 18.6 | 24.9 | 30.7 |
| US&J/E (9) | 22.8 | 35.3 | 57.6 |
| E/E (13) | 22.8 | 35.5 | 55.7 |
| NIC (11) | 25.7 | 41 | 78.7 |

(Sample Size)

*Note:* Volume producers include the American "Big Three", Fiat, PSA, Renault, and Volkswagen in Europe, and all of the companies from Japan.

- J/J = Japanese-owned plants in Japan.
- J/NA = Japanese-owned plants in North America, including joint venture plants with American firms.
- US/NA = American-owned plants in North America.
- US&J/E = American- and Japanese-owned plants in Europe.
- E/E = European-owned plants in Europe.
- NIC = Plants in newly industrializing countries: Mexico, Brazil, Taiwan, and Korea.

*Source:* IMVP World Assembly Plant Survey

---

What we find instead is that there is a considerable range of productivity performance in Japan, indeed a difference of two to one between the best plant and the worst in both productivity and quality. The differences along other dimensions—space utilization, level of inventories, percentage of the factory devoted to rework area—are much less, but there is still variation.

In North America, we quickly discovered that Framingham was in fact the worst American-owned plant. Average Big Three performance in late 1989 was much better—48 percent more effort and 50 percent more defects, compared with the Framingham/Takaoka gap in 1986 of nearly twice the effort and three

#### FIGURE 4.4

**Assembly Plant Quality, Volume Producers, 1989**

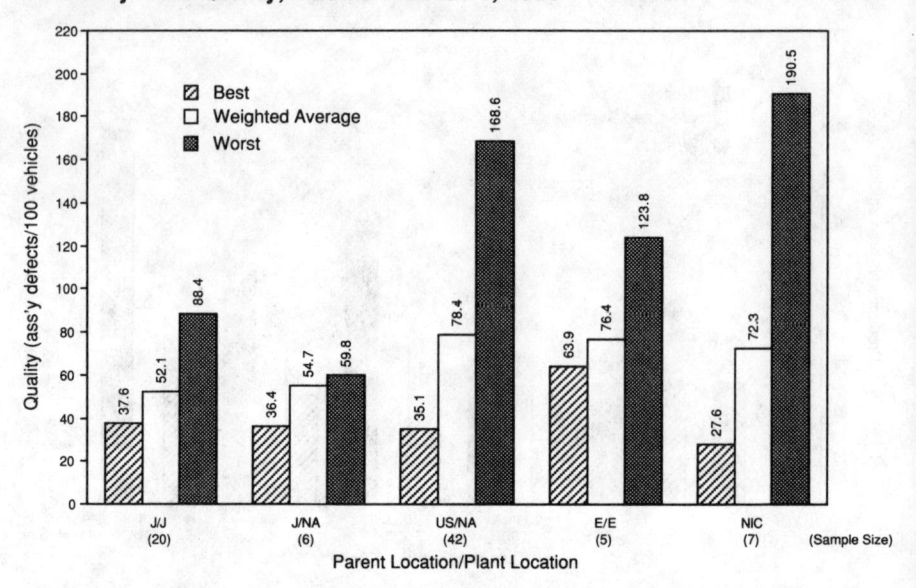

| Parent Location/Plant Location | Best | Weighted Average | Worst |
|---|---|---|---|
| J/J (20) | 37.6 | 52.1 | 88.4 |
| J/NA (6) | 36.4 | 54.7 | 59.8 |
| US/NA (42) | 35.1 | 78.4 | 168.6 |
| E/E (5) | 63.9 | 76.4 | 123.8 |
| NIC (7) | 27.6 | 72.3 | 190.5 |

Quality (ass'y defects/100 vehicles)

*Note:* Quality is expressed as the number of defects per 100 cars traceable to the assembly plant, as reported by owners in the first three months of use. The reports only include cars sold in the United States.

*Source:* IMVP World Assembly Plant Survey, utilizing a special tabulation of defects by assembly plant provided by J. D. Power and Associates.

---

times the defects. Even more striking, Ford, the originator of mass production seventy-five years ago, is now practically as lean in its North American assembly plants as the average Japanese transplant in North America.[5] The best U.S.-owned plants in North America are now nearly as productive as the average Japanese plant—and are very nearly equal in quality.

Perhaps most striking was our finding about Europe. Framingham, the North American plant that fared so poorly in comparison with Takaoka and which has now been closed, in fact had considerably better productivity in 1986 than the average European plant had achieved by 1989. Indeed, as we marched through plant after plant we came to a remarkable conclusion: Europe, once the cradle of craft production in the motor industry, is now truly the home of classic mass production. Average American performance—under unrelenting pressure from the Japanese transplants in North America—has improved dramatically, partly by closing the worst plants, such as Framingham, and partly by adopting lean production techniques at others. Europe, by contrast, has not yet begun to close the competitive gap.

Regarding the Japanese transplants in North America, we found about what we expected. Their average performance is about comparable to the average Japanese plant in terms of quality but lags about 25 percent in terms of productivity. We believe these differences are partly due to the fact that the transplants are still at an early point in the learning curve with respect to lean production. The differences are also due to different methods of obtaining supplies that necessitate extra work, a point we will return to in Chapter 6.[6]

However, there is important variation among the transplants as well. For example, one of the transplants has the least efficient utilization of manufacturing space in the entire world sample. In general, we found that the best-performing companies in Japan run the best-performing transplants in North America, suggesting that most of the variation observed is due to differences in management.

Finally, the assembly plants in the developing countries, notably Brazil, Korea, Mexico, and Taiwan, show an extraordinary range of performance. The best plant in terms of quality, Ford at Hermosillo, Mexico, in fact had the best assembly-plant quality in the entire volume plant sample, better than that of the best Japanese plants and the best North American transplants. The best developing country plant was also surprisingly efficient, particularly given its modest level of automation. By contrast, the worst developing country plants were very poor performers, with poor quality and abysmal productivity.

What accounts for the difference? We believe it can be traced to the assembly of a product from a lean-development process (as at Hermosillo, where the car assembled was a variant of a Mazda 323) and doing so with management assistance from a firm mastering lean production. (In the case of Hermosillo, this was directly from Ford, but in several other cases an independent firm had received significant and continuing Japanese management assistance, effectively becoming a transplant).

These findings require a dramatic reordering of our mental map of the industrial world, which we believe many readers will find very difficult: We must now stop equating "Japanese" with "lean" production and "Western" with "mass" production. In fact, some plants in Japan are not particularly lean, and a number of Japanese-owned plants in North America are now demonstrating that lean production can be practiced far away from Japan. At the same time, the best American-owned plants in North America show that lean production can be implemented fully by Western companies, and the best plants in the developing countries show that lean production can be introduced anywhere in the world.

## THE STRANGE CASE OF THE "CRAFT" PRODUCERS

The productivity and quality data in figures 4.3 and 4.4 are only for mass-market cars, that is, Fords but not Lincolns, Toyotas but not Lexus, Volkswagens but not Mercedes. From the outset, we believed that assembly plants are all pretty much the same in what they actually do, no matter how prestigious the brand they're putting together. The same type of robots, indeed often identical models from the same manufacturer, make both the Volkswagen and the Mercedes body welds. Paint is applied in practically identical paint booths, and final assembly involves the installation, largely by hand, of thousands of parts as the vehicle moves along a lengthy assembly line. The real difference between the mass-market and the luxury car is that the latter may have a thicker gauge of steel in its body, extra coats of paint, thicker insulation, and many more luxury add-on features.

While obvious to us, this idea is not universally accepted even in the auto industry and is certainly not the view of the broader public. Repeatedly, executives told us that our productivity and quality findings might be correct for the average car and light truck, but "luxury cars are different."

We set out to find out for certain by conducting a special world survey of assembly plants making luxury cars. We went to the Japanese large-car plant that we believe, based on our survey of the same company's mass-market car plants, to be the best in the world. In North America, we looked at the Lincoln and Cadillac plants. In Europe we visited Audi, BMW, Mercedes, Volvo, Rover, Saab, and Jaguar. In each case, we carefully standardized the tasks being undertaken and the specifications of the vehicle, so that we were in fact asking how much effort each plant would need to perform standard assembly steps on the smaller and less elaborate standard car and how many errors it would make in the process. So the actual amount of effort expended in each plant is actually much greater than that shown in figures 4.5 and 4.6. In addition, we adjusted for absenteeism, which runs at 25 percent in many of these European plants compared with 5 percent or less in Japan. The hours in our table represent hours actually worked, not hours on the payroll.

Our findings were eye-opening. The Japanese plant requires one-half the effort of the American luxury-car plants, half the effort of the best European plant, a quarter of the effort of the average European plant, and one-sixth the effort of the worst European luxury-car producer. At the same time, the Japanese plant greatly exceeds the quality level of all plants except one in Europe—and this European plant requires four times the effort of the Japanese plant to assemble a comparable product. No

**FIGURE 4.5**

**Luxury Car Assembly Plant Productivity, 1989**

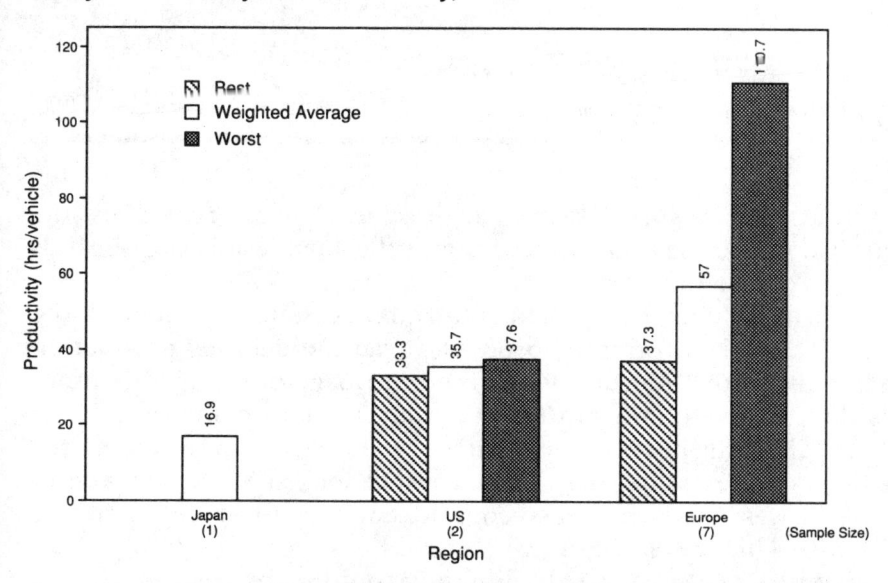

| Region | Best | Weighted Average | Worst |
|---|---|---|---|
| Japan (1) | | 16.9 | |
| US (2) | 33.3 | 35.7 | 37.6 |
| Europe (7) | 37.3 | 57 | 110.7 |

Productivity (hrs/vehicle)

*Note:* "Luxury cars" include those produced by the European "specialist producers"—Daimler-Benz, BMW, Volvo, Saab, Rover, Jaguar, Audi, and Alfa Romeo—and by Cadillac and Lincoln in North America. The Japanese luxury category includes the Honda Legend, the Toyota Cressida, and the Mazda 929, the three most expensive sedans being built by the Japanese companies for export in 1989. The Toyota Lexus and Nissan Infiniti models were too recent to include.

*Source:* IMVP World Assembly Plant Survey

## FIGURE 4.6

### Luxury Car Assembly Plant Quality, 1989

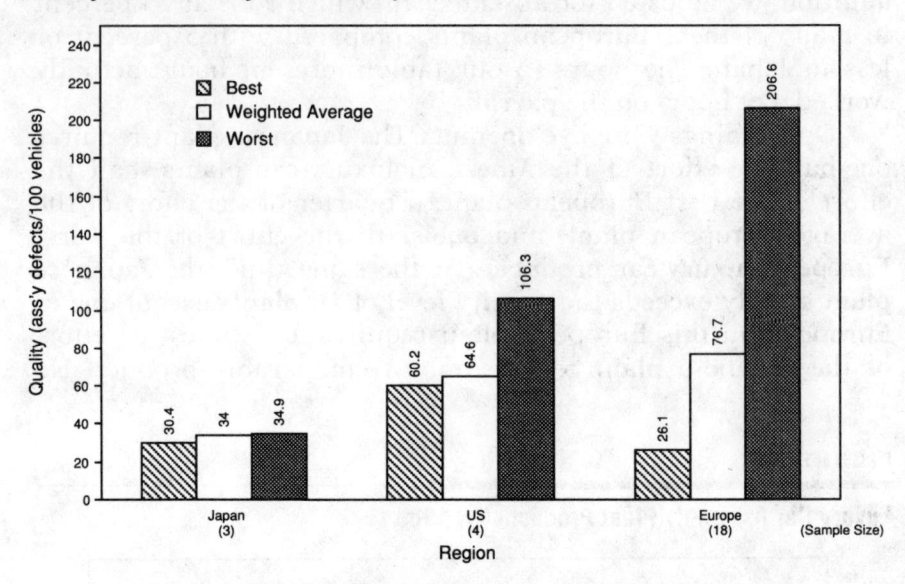

[Bar chart — Quality (ass'y defects/100 vehicles) by Region, showing Best, Weighted Average, and Worst values:
- Japan (3): 30.4, 34, 34.9
- US (4): 60.2, 64.6, 106.3
- Europe (18): 26.1, 76.7, 206.6]

*Note:* "Luxury cars" are as defined in Figure 4.5.
*Source:* IMVP World Assembly Plant Survey, based on data supplied by J. D. Power and Associates.

wonder the Western luxury-car producers are terrified by the arrival of Lexus, Infiniti, Acura, and the Japanese luxury brands still to come.

In reviewing these data, many readers may wonder if the difference lies in greater product variety and lower production scale in Europe. Certainly our mental image of these companies is that of low-volume craft production. In fact, this is simply not true. The European plants, with one exception, produce at the same volume as the mass-producers we looked at earlier, and in most cases produce a *less* complex mix of products than the Japanese luxury car plant we surveyed.

When we visited the high-quality but low-productivity European plant we just mentioned, we didn't have to go far to find the basic problem: a widespread conviction among managers and workers that they were craftsmen. At the end of the assembly line was an enormous rework and rectification area where armies of technicians in white laboratory jackets labored to bring the finished vehicles up to the company's fabled quality standard. We found that a third of the total effort involved in assembly occurred in this area. In other words, the German plant was expending more effort to fix the problems it had just created than the Japanese plant required to make a nearly perfect car the first time.

We politely inquired of these white-smocked workers exactly what they were doing. "We're craftsmen, proof of our company's dedication to quality," they replied. These "craftsmen" would have been surprised to learn that they were actually doing the work of Henry Ford's fitters in 1905—adjusting off-standard parts, fine-tuning parts designed so as to need adjustment, and rectifying incorrect previous assembly work so that everything would work properly in the end.

Certainly, these workers are highly skilled and the work they do is no doubt challenging, since every problem is different. However, from the standpoint of the lean producer this is pure *muda*—waste. Its cause: the failure to design easy-to-assemble parts and failure to track down defects as soon as they are discovered so that they never recur. When employees don't take this important last step, subsequent assembly work compounds the initial problem and it's necessary to call for the craftsman to put things right.

Our advice to any company practicing "craftsmanship" of this sort in any manufacturing activity, automotive or otherwise, is simple and emphatic: Stamp it out. Institute lean production as quickly as possible and eliminate the need for all craftsmanship at the source. Otherwise lean competitors will overwhelm you in the 1990s.

## THE IMVP WORLD ASSEMBLY PLANT SURVEY IN SUMMARY

Figure 4.7 summarizes a number of dimensions of current worldwide performance of the volume producers at the assembly plant level in addition to productivity and quality. In particular, it is striking to note the difference between average Japanese performance and the average in North America and Europe in terms of the size of repair area needed, the fraction of workers organized into teams, the number of suggestions received (and the lack so far of suggestion systems in the Japanese transplants), and the amount of training given new assembly workers.

#### FIGURE 4.7

**Summary of Assembly Plant Characteristics, Volume Producers, 1989 (Averages for Plants in Each Region)**

|  | Japanese in Japan | Japanese in North America | American in North America | All Europe |
|---|---|---|---|---|
| *Performance:* | | | | |
| Productivity (hours/veh.) | 16.8 | 21.2 | 25.1 | 36.2 |
| Quality (assembly defects/100 vehicles) | 60.0 | 65.0 | 82.3 | 97.0 |
| *Layout:* | | | | |
| Space (sq. ft./vehicle/year) | 5.7 | 9.1 | 7.8 | 7.8 |
| Size of Repair Area (as % of assembly space) | 4.1 | 4.9 | 12.9 | 14.4 |
| Inventories (days for 8 sample parts) | .2 | 1.6 | 2.9 | 2.0 |
| *Work Force:* | | | | |
| % of Work Force in Teams | 69.3 | 71.3 | 17.3 | .6 |
| Job Rotation (0 = none, 4 = frequent) | 3.0 | 2.7 | .9 | 1.9 |
| Suggestions/Employee | 61.6 | 1.4 | .4 | .4 |
| Number of Job Classes | 11.9 | 8.7 | 67.1 | 14.8 |
| Training of New Production Workers (hours) | 380.3 | 370.0 | 46.4 | 173.3 |
| Absenteeism | 5.0 | 4.8 | 11.7 | 12.1 |
| *Automation:* | | | | |
| Welding (% of direct steps) | 86.2 | 85.0 | 76.2 | 76.6 |
| Painting (% of direct steps) | 54.6 | 40.7 | 33.6 | 38.2 |
| Assembly (% of direct steps) | 1.7 | 1.1 | 1.2 | 3.1 |

*Source:* IMVP World Assembly Plant Survey, 1989, and J. D. Power Initial Quality Survey, 1989

One additional and very important finding of the survey bears note: the relation between productivity and quality. When we first began the survey and correlated productivity with quality in all plants, we found almost no relationship. What's more, this did not change over time. In Figure 4.8, showing the relationship across the world at the end of 1989, the correlation between productivity and quality is .15.

This seemed puzzling. We thought it should either be negatively correlated—plants with high quality should require more effort to achieve this, as Western factory managers had long thought—or it should be positively correlated—quality should be "free," as many writers on Japanese manufacturing had suggested. The answer to the puzzle, as a moment's examination of

**FIGURE 4.8**

**Productivity versus Quality in the Assembly Plant, Volume Producers, 1989**

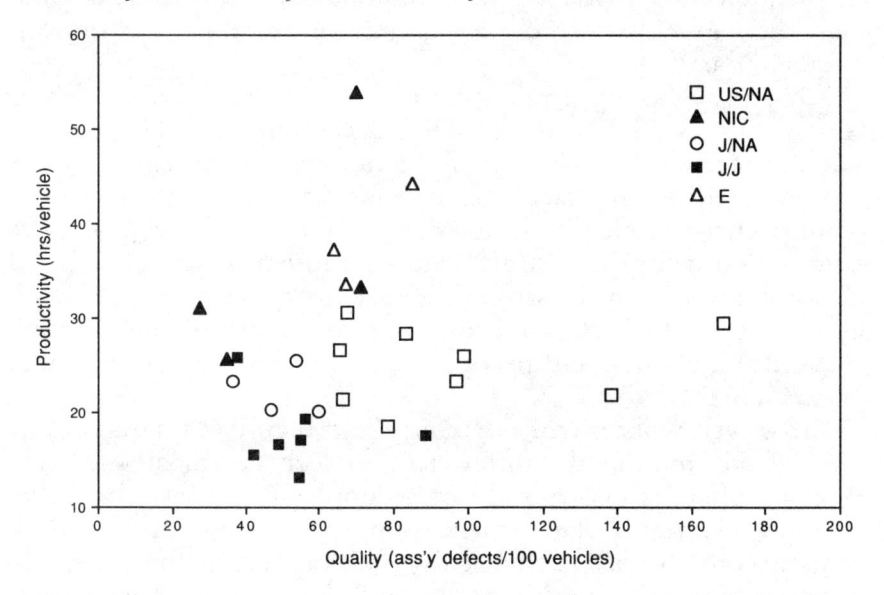

[Scatter plot showing Productivity (hrs/vehicle) on y-axis from 10 to 60, versus Quality (ass'y defects/100 vehicles) on x-axis from 0 to 200. Legend: □ US/NA, ▲ NIC, ○ J/NA, ■ J/J, △ E]

*Source:* IMVP World Assembly Plant Survey, 1989

Figure 4.8 will show, is that both trends are in evidence and they cancel each other out. The Japanese domestic and transplant facilities are all concentrated in the lower left corner of the figure. For these lean plants, quality really is free. Removing these plants leaves a pattern in which plants tend to have high quality or high productivity but not both. For these mass-producers, quality is expensive when it can be achieved at all.

## GETTING TO LEAN

We have periodically reviewed our survey findings with practically all the world's motor-vehicle producers, the main sponsors of the IMVP. So the figures we report here don't come as a surprise to these companies and are now generally accepted as an accurate summary of the general state of competition at the factory level.

However, determining who stands where in world competition differs from explaining precisely what the also-rans need to do to catch up. As we have reviewed our data with these companies, their executives and managers have questioned us on four points in particular.

First, they ask whether automation is the secret. Our answer is that it is and it isn't. Figure 4.9 shows the relation between the fraction of assembly steps that are automated—either by robotics or more traditional "hard" automation—and the productivity of plants. There is clearly a downward slope to the right—more automation means less effort. (Stated another way, higher levels of automation show a strong negative correlation ($-.67$) with higher levels of effort.) We estimate that on average automation accounts for about one-third of the total difference in productivity between plants.

However, what is truly striking about Figure 4.9 is that at any level of automation the difference between the most and least efficient plant is enormous. For example, the least automated Japanese domestic plant in the sample (with 34 percent of all steps accomplished automatically), which is also the most efficient plant in the world, needs half the human effort of one comparably automated European plant and a third the effort of another. Looking farther to the right in Figure 4.9, we can see that the European plant that is the most automated in the world (with 48 percent of all assembly steps done by automation) requires 70 percent more effort to perform our standard set of assembly tasks on our standard car than is needed by the most efficient plant with only 34 percent automation.

The obvious question is, how can this be? From our survey findings and plant tours, we've concluded that high-tech plants that are improperly organized end up adding about as many indirect technical and service workers as they remove unskilled direct workers from manual assembly tasks.

What's more, they have a hard time maintaining high yield, because breakdowns in the complex machinery reduce the fraction of the total operating time that a plant is actually producing vehicles. From observing advanced robotics technology in many plants, we've devised the simple axiom that lean organization must come before high-tech process automation if a company is to gain the full benefit.[7]

The also-rans' second question is, Does the manufacturability (ease of assembly) of the product make the difference, rather than

## FIGURE 4.9

**Automation versus Productivity, Volume Producers, 1989**

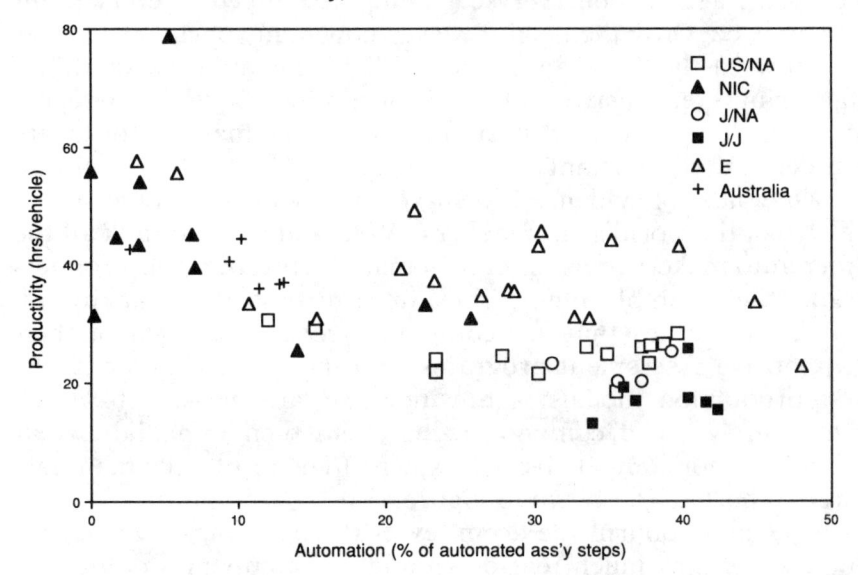

*Note:* "Automation" equals the percent of assembly tasks that have been automated. Automation includes both fixed automation such as multi-welders and flexible automation using robots. Automation of materials handling is not included.

*Source:* IMVP World Assembly Plant Survey, 1989

the operation of the factory? Understandably, union leaders have often asked us this question as well. Donald Ephlin, now retired from his position as vice-president of the United Auto Workers in the United States, engaged us in a dialogue on this point throughout the life of the IMVP.

How much of the competitive gap between good firms and bad, he wanted to know, lies with the unionized workers in the plant and how much with engineers and managers far away in the corporate development offices. His argument has been simple: "The workers I represent in American plants are getting the blame for problems they are helpless to correct." Ephlin argued that putting in place organizational improvements—just-in-time inventory, a cord that allowed the worker to stop the line, and so forth—would make a difference, but that none of those improvements could make a plant fully competitive if the product design was defective.

Answering the manufacturability question definitively is difficult, because we would need to perform what auto makers call a tear-down analysis on every car being assembled in every plant we surveyed. Only then could we see how many parts the car has and how easily they can be assembled. This analysis would be staggeringly expensive and time-consuming. So we can report only some interesting but partial evidence that manufacturability is indeed very important.

One piece of evidence is a survey we conducted in the spring of 1990 of the world's auto makers.[8] We asked them to rank all the other auto makers in terms of how manufacturable their products are at the assembly plant. They were to base their ranking on tear-down studies that car companies conduct as part of their competitive assessment programs. (Strange as it may seem, the first production models of any new car are unlikely to reach consumers. Instead, competitors buy them, then immediately tear them apart for competitive assessment.) The results the manufacturers reported are shown in Figure 4.10.

We can't confirm the accuracy of these findings, because we don't know how much tear-down analysis companies do or how well they do it. When we began our assembly-plant survey, we were amazed to discover that very few car companies conducted systematic benchmarking studies of their competitors. Nevertheless, the companies responding were in close agreement on which producers design the most manufacturable designs, and the findings correlate nicely with company performance on our productivity and quality indices. This suggests that manufacturability is conducive to high performance in the factory.

Further evidence comes from a recent comparison General Motors made between its new assembly plant at Fairfax, Kansas, which makes the Pontiac Grand Prix version of its GM-10 model, and Ford's assembly plant for its Taurus and Mercury Sable models near Atlanta. This comparison was based on tearing down both cars, then using shop manuals to reconstruct the assembly process.

GM found a large productivity gap between its plant and the Ford plant—both make cars in the same size class, with similar levels of optional equipment and selling in the same market segment. After careful investigation, GM concluded that 41 percent of the productivity gap could be traced to the manufacturability of the two designs, as shown in Figure 4.11. For example, the Ford car has many fewer parts—ten in its front bumper

RUNNING THE FACTORY 97

**FIGURE 4.10**

**Manufacturability of Products in the Assembly Plant, Producers Ranked by Other Producers, 1990**

| Producer | Average Rank | Range of Rankings |
|---|---|---|
| Toyota | 2.2 | 1–3 |
| Honda | 3.9 | 1–8 |
| Mazda | 4.8 | 3–6 |
| Fiat | 5.3 | 2–11 |
| Nissan | 5.4 | 4–7 |
| Ford | 5.6 | 2–8 |
| Volkswagen | 6.4 | 3–9 |
| Mitsubishi | 6.6 | 2–10 |
| Suzuki | 8.7 | 5–11 |
| General Motors | 10.2 | 7–13 |
| Hyundai | 11.3 | 9–13 |
| Renault | 12.7 | 10–15 |
| Chrysler | 13.5 | 9–17 |
| BMW | 13.9 | 12–17 |
| Volvo | 13.9 | 10–17 |
| PSA | 14.0 | 11–16 |
| Saab | 16.4 | 13–18 |
| Daimler-Benz | 16.6 | 14–18 |
| Jaguar | 18.6 | 17–19 |

Note: These rankings were compiled by summing responses to a survey of the nineteen major assembler firms. Eight firms returned the survey in usable form—two American, four European, one Japanese, and one Korean. The firms were asked to rank all nineteen firms "according to how good you think each company is at designing products that are easy for an assembly plant to build."

Source: IMVP Manufacturability Survey, 1990

**FIGURE 4.11**

**Ford Atlanta Assembly Plant versus GM Fairfax Assembly Plant, 1989**

Productivity Difference, Allocated by Cause:
| | |
|---|---|
| Sourcing | 9% |
| Processing | 2% |
| Design for Manufacture | 41% |
| Factory Practice | 48% |
| | 100% |

Source: General Motors

compared with 100 in the GM Pontiac—and the Ford parts fit together more easily. (The other major cause of the productivity gap was plant organizational practices of the type we have just discussed. The GM study found that the level of automation—which was actually much higher in the GM plant—was not a factor in explaining the productivity gap.)

Ease of manufacture is not an accident. Rather, it's one of the most important results of a lean-design process. We'll look at this point more carefully in Chapter 5.

A third question that often crops up when we review our survey findings with companies is product variety and complexity. The factory manager we encountered in Chapter 3, who maintained he could compete with anyone if he could only focus his factory on a single standardized product, is typical of many Western managers. This is certainly an interesting idea and it has a simple logic to commend it.

However, in our survey we could find no correlation at all between the number of models and body styles being run down a production line and either productivity or product quality. We tried a different approach by comparing what was being built in plants around the world in terms of "under-the-skin" complexity. This was a composite measure composed of the number of main body wire harnesses, exterior paint colors, and engine/transmission combinations being installed on a production line, plus the number of different parts being installed and the number of different suppliers to an assembly plant. The results were even less assuring to those thinking that a focused factory is the solution to their competitive problems: The plants in our survey with the highest under-the-skin complexity also had the highest productivity and quality. These of course were the Japanese plants in Japan.[9]

## LEAN ORGANIZATION AT THE PLANT LEVEL

Those company executives, plant managers, and union leaders accepting our conclusion that automation and manufacturability are both important to high performance plants, but that gaining the full potential of either requires superior plant management, usually raise a final question that we find most interesting: What are the truly important organizational features of a lean plant—

the specific aspects of plant operations that account for up to half of the overall performance difference among plants across the world? And how can these be introduced?

The truly lean plant has two key organizational features: *It transfers the maximum number of tasks and responsibilities to those workers actually adding value to the car on the line, and it has in place a system for detecting defects that quickly traces every problem, once discovered, to its ultimate cause.*

This, in turn, means teamwork among line workers and a simple but comprehensive information display system that makes it possible for everyone in the plant to respond quickly to problems and to understand the plant's overall situation. In old-fashioned mass-production plants, managers jealously guard information about conditions in the plant, thinking this knowledge is the key to their power. In a lean plant, such as Takaoka, all information—daily production targets, cars produced so far that day, equipment breakdowns, personnel shortages, overtime requirements, and so forth—are displayed on *andon* boards (lighted electronic displays) that are visible from every work station. Every time anything goes wrong anywhere in the plant, any employee who knows how to help runs to lend a hand.

So in the end, it is the dynamic work team that emerges as the heart of the lean factory. Building these efficient teams is not simple. First, workers need to be taught a wide variety of skills—in fact, all the jobs in their work group so that tasks can be rotated and workers can fill in for each other. Workers then need to acquire many additional skills: simple machine repair, quality-checking, housekeeping, and materials-ordering. Then they need encouragement to think actively, indeed *pro*actively, so they can devise solutions before problems become serious.

Our studies of plants trying to adopt lean production reveal that workers respond only when there exists some sense of reciprocal obligation, a sense that management actually values skilled workers, will make sacrifices to retain them, and is willing to delegate responsibility to the team. Merely changing the organization chart to show "teams" and introducing quality circles to find ways to improve production processes are unlikely to make much difference.

This simple fact was brought home to us by one of our early studies of Ford and General Motors plants in the United States. In the Ford plants we found that the basic union-management contract had not been changed since 1938, when Ford was finally forced to sign a job control contract with the UAW. Workers continued to have narrow job assignments and no formal team structure was in place. Yet, as we walked through plant after plant we observed that teamwork was actually alive and well. Workers were ignoring the technical details of the contract on a massive scale in order to cooperate and get the job done.[10]

By contrast, in a number of General Motors showcase plants we found a new team contract in place and all the formal apparatus of lean production. Yet, a few moments' observation revealed that very little teamwork was taking place and that morale on the plant floor was very low.

How do we account for these seeming contradictions? The answer is simple. The workers in the Ford plants had great confidence in the operating management, who worked very hard in the early 1980s to understand the principles of lean production. They also strongly believed that if all employees worked together to get the job done in the best way the company could protect their jobs. At the GM plants, by contrast, we found that workers had very little confidence that management knew how to manage lean production. No wonder, since GM's focus in the early 1980s was on devising advanced technology to get rid of the workers. The GM workers also had a fatalistic sense that many plants were doomed anyway. In these circumstances, it is hardly surprising that a commitment to lean production from the top levels of the corporation and the union had never translated into progress on the plant floor.

We'll return to the thorny question of how lean production can be introduced into existing mass-production factories in Chapter 9.

## IS LEAN PRODUCTION HUMANLY FULFILLING?

As we noted in Chapter 2, Henry Ford's sword was double-edged. Mass production made mass consumption possible, while it made factory work barren. Does lean production restore the satisfaction of work while raising living standards, or is it a sword even more double-edged than Ford's?

Opinions are certainly divided. Two members of the United Automobile Workers Union in the United States have recently argued that lean production is even worse for the worker than mass production.[11] They go so far as to label the lean-production system instituted at NUMMI in California management by stress, because managers continually try to identify slack in the system—unused work time, excess workers, excess inventories—and remove them. Critics argue that this approach makes *Modern Times* look like a picnic. In Charlie Chaplin's widget factory at least the workers didn't have to think about what they were doing and try to improve it.

A second critique of lean production comes in the form of what might be called "neocraftsmanship." This has been put in operation in only a few plants in Sweden, but it draws wide attention across the world because it appeals to a seemingly unshakable public faith in craftsmanship.

Let's take Volvo's new Udevalla plant in western Sweden as an example. At Udevalla, teams of Volvo workers assemble Volvo's 740 and 760 models on stationary assembly platforms in small work cells. Each team of ten workers is responsible for putting together an entire vehicle from the point it emerges from the paint oven. Looked at in one way, this system is a return full circle to Henry Ford's assembly hall of 1903, which we and the rest of the world left behind in Chapter 2. The cycle time—the interval before the worker begins to repeat his or her actions—increases at Udevalla to several hours, from about one minute in a mass- or lean-production assembly plant. In addition, workers in the assembly team can set their own pace, so long as they complete four cars each day. They can also rotate jobs within the teams as they desire. Automated materials-handling delivers the parts needed for each car to the work team. Proponents of the Udevalla system argue that it can match the efficiency of lean-production plants while providing a working environment that is much more humane.

We strongly disagree with both points. We believe that a vital, but often misunderstood, difference exists between tension and a continuing challenge and between neocraftsmanship and lean production.

To take the first point, we agree that a properly organized lean-production system does indeed remove all slack—that's why it's *lean*. But it also provides workers with the skills they need to control their work environment and the continuing challenge of making the work go more smoothly. While the mass-production plant is often filled with mind-numbing stress, as workers struggle to assemble unmanufacturable products and have no way to improve their working environment, lean production offers a creative tension in which workers have many ways to address challenges. This creative tension involved in solving complex problems is precisely what has separated manual factory work from professional "think" work in the age of mass production.

To make this system work, of course, management must offer its full support to the factory work force and, when the auto market slumps, make the sacrifices to ensure job security that has historically been offered only to valued professionals. It truly is a system of reciprocal obligation.

What's more, we believe that once lean production principles are fully instituted, companies will be able to move rapidly in the 1990s to automate most of the remaining repetitive tasks in auto assembly—and more. Thus by the end of the century we expect that lean-assembly plants will be populated almost entirely by highly skilled problem solvers whose task will be to think continually of ways to make the system run more smoothly and productively.

The great flaw of neocraftsmanship is that it will never reach this goal, since it aspires to go in the other direction, back toward an era of handcrafting as an end in itself.

We are very skeptical that this form of organization can ever be as challenging or fulfilling as lean production. Simply bolting and screwing together a large number of parts in a long cycle rather than a small number in a short cycle is a very limited vision of job enrichment. The real satisfaction presumably comes in reworking and adjusting every little part so that it fits properly. In the properly organized lean-production system, this activity is totally unnecessary.

Finally, the productivity of the Udevalla system is almost certain to be uncompetitive even with mass production, much less lean production. We have not audited Udevalla or Kalmar, the two Volvo plants operated on the neocraft model, but some simple arithmetic suggests that if ten workers require 8 hours simply to assemble four vehicles (not including welding the body, painting it, and gathering necessary materials)—for a total of 20 assembly hours per vehicle—Udevalla can hardly hope to compete with our survey's leading lean-production plant, which requires only 13.3 hours to weld, paint, and assemble a slightly smaller and less elaborate vehicle.

Before leaving this point, we offer one final reason why lean production is unlikely to prove more oppressive than mass production. Simply put, lean production is *fragile*. Mass production is designed with buffers everywhere—extra inventory, extra space, extra workers—in order to make it function. Even when parts don't arrive on time or many workers call in sick or other workers fail to detect a problem before the product is mass-produced, the system still runs.

However, to make a lean system with no slack—no safety net—work at all, it is essential that every worker try very hard. Simply going through the motions of mass production with one's head down and mind elsewhere quickly leads to disaster with lean production. So if management fails to lead and the work force feels that no reciprocal obligations are in force, it is quite predictable that lean production will revert to mass production. As one lean-production manager remarked during a plant tour: "Mass production is simply lean production run by the rule book, so that no one takes initiative and responsibility to continually improve the system."

This last point raises some profound questions about the spread of lean production across the whole world, a topic that occupies our attention in Chapter 9. However, at this point, we need to follow the logic of lean production from the assembly plant back to product development. As we'll see, the nature of the modern motorcar—a highly complex product with more than 10,000 parts—requires a highly complex design and engineering system. And, as in every other aspect of production, the lean approach to coordinating this system is fundamentally different from that of mass production.

# 5

## DESIGNING THE CAR

### GM-10: PRODUCT DEVELOPMENT IN A MASS-PRODUCTION FIRM

In 1981 General Motors began to plan a replacement for its just launched front-drive A-cars and its older rear-drive G-cars, the company's offerings in the intermediate-size segment of the North American market. The A-cars—the Chevrolet Celebrity, Pontiac 6000, Oldsmobile Ciera, and Buick Century—would normally have remained in production for ten years. However, GM knew that Ford was developing a new intermediate-size model for introduction in 1985, and the Japanese companies were thought to be planning a much stronger presence in this segment. (Intermediate is one of the four standard size categories traditionally used to segment the American car market: subcompact, compact, intermediate, and full size.)

The intermediate-size segment of the market had by then become GM's volume base, accounting for about one-third of the company's annual car sales in North America. Senior executives at GM concluded that they dare wait no longer than 1986 for a new model. They knew that if they tried to run the usual ten-year cycle with A bodies while continuing with the older G bodies, they would fall badly behind Ford and the Japanese. So they set

---

This chapter is based on the research of Takahiro Fujimoto, Andrew Graves, Kentaro Nobeoka, and Antony Sheriff.

# DESIGNING THE CAR

in motion the extraordinarily complex and expensive process of developing a new car.

All large automobile companies—mass production or lean—face the same basic problem in developing a new product. A number of functional departments—marketing, power train engineering, body engineering, chassis engineering, process engineering, and factory operations—must collaborate intensively over an extended period of time to develop the new car successfully. The question is how.

The simplest approach would be to create a totally self-contained project team consisting of the necessary number of planners and engineers. A team manager could orchestrate the efforts of this group over a period of years until the project was completed.

In fact, no company in the world, mass-producer or lean, does this. The reasons are simple. Every company has a range of models, mechanical components, and factories that must be shared. Model A will share a transmission with Model B and be built alongside Model C in the same factory. Isolating the transmission engineers and factory managers for Model A in a self-sufficient team would never work, since their efforts would soon be at cross-purposes with teams working on models B and C. Isolating the product planners would never work either, because their designs might overlap with other new products in the planning pipeline. Moreover, engineers working in isolation would soon loose contact with the technical frontier of their specialty, as it is pushed ahead by research activities in the functional departments. The result: Their designs would not be state-of-the-art.

In consequence, most automotive companies develop some sort of matrix in which every employee involved in developing a product reports both to a functional department and to a development program. The leadership challenge is managing the matrix to satisfy the needs of both the functional department and the product-development program.

At General Motors, meeting this challenge has been particularly critical. From the 1930s to the end of the 1950s, the company turned out five basic models—the Chevrolet, Pontiac, Oldsmobile, Buick, and Cadillac. The five had separate chassis, bodies, and engines but shared hundreds, or even thousands, of other parts—pumps, electrical components, springs, bearings, and glass. So the development of a new model by any of the car divisions involved a complex interaction with the other car divisions and the components divisions that made the shared parts. This was the organizational consequence of Alfred Sloan's determination to share as many parts as possible to gain economies of scale.

After 1959, when GM introduced its first small cars, the situation became still more complex. By the late 1960s the company was offering four separate sizes of car in each car division, except Cadillac, where it offered two models. To preserve economies of scale while doing this, GM began to share a basic model between its divisions, giving the model sold under each divisional name a slightly different appearance. So the new intermediate model introduced in 1968 was served up as a Chevrolet Chevelle, a Pontiac Tempest, an Olds F-85, and a Buick Skylark. These cars had different exterior sheet metal and different dashboards and door trims on the interiors but used exactly the same basic components, including engines and chassis, under their metal skins. In other words, everything tucked out of sight was exactly the same. To develop these products, the company now had to coordinate the needs of four marketing divisions as well, each wanting a different character—sporty, conservative, technologically advanced, luxurious—in their version to satisfy the expectations of traditional buyers of that division's cars.

GM's approach to developing its new model, known inside the company as the GM-10, was standard. Senior executives designated a program manager to take the lead in coordinating the functional departments involved in the development process. Robert Dorn, chief engineer of GM's Pontiac Division, was selected as the GM-10 manager and given a $7-billion development budget. He set up an office in GM's Chevrolet Division, recruited a small personal staff, and went to work. (Since the GM system doesn't have an Office of Program Manager, those selected for this role are, in effect, nomads and must find a division to take them in.)

His first step was to get the four car divisions to agree on the target market of consumers for the car and the product features these buyers would find most appealing. To get the job done, he ordered large amounts of survey research among consumers and an analysis of sales patterns in the market.

Key decisions made during this process included the physical dimensions of the new cars, their general appearance and performance, the target market and price (around $14,000) and accompanying target cost, fuel economy (about 24 miles per gallon), and body styles. All four divisions wanted a two-door coupe and a four-door sedan, and several requested a station wagon.

Dorn's group took this information and consulted with the GM Styling Center on the exact external and internal appearance of each model. This process begins with rough sketches, proceeds to detailed models in clay, then advances to actual prototypes, which are shown to representative potential buyers for their reactions.

When all the thousands of decisions on specifications, appearance, and performance had been made, Dorn's group conveyed the details to the next group of specialists in what was then General Motors' Fisher Body Division and the components-engineering divisions. There engineers worked out the precise specifications of every major part and, more important, decided which parts could be carried over from the existing A cars and which could be obtained from other GM products. Parts that could neither be carried over nor shared had to be designed from scratch. (This advanced engineering is the most time-consuming and expensive part of any development program and needs to start as soon as possible.)

By this point, Robert Dorn was becoming concerned. The GM-10 program was steadily slipping its five-year timetable, and Dorn's small staff seemed powerless to make it go faster. Much of the problem stemmed from the fact that Dorn and his group were in fact coordinators, rather than managers. In other words, they reasoned with people in order to coordinate efforts; they weren't leaders who gave orders and expected to be obeyed. When they urged the engineering departments to go faster, they were met with promises but little action. Clearly, in this matrix, each employee was more concerned with pleasing his or her functional department boss than the GM-10 program coordinator. For example, if the coordinator pointed out that a feature of the engine needed changing so it would run correctly, the team representative from power train engineering would stall, knowing that the engine was just fine for the cars that accounted for most of GM's production volume.

As the program fell further behind, its problems multiplied. The erosion of the station wagon segment by minivans caused cancellation of the station wagon version of GM-10, and the introduction of the Ford Taurus in 1985 caused GM to redesign the exterior sheet metal of the GM-10 cars because senior executives felt they would otherwise be too similar to the Ford products.

Finally, in 1985, Dorn had enough and resigned. He was replaced by Gary Dickenson, who faced the next major hurdle in the GM-10 program—moving the completed product design from Fisher Body and the components-engineering departments over to the then General Motors Assembly Division (GMAD), which was charged with actually manufacturing the cars. GMAD was a monolithic organization (now disbanded in a major reorganization), with its own internal culture and career path, and Dickenson soon became as frustrated as Dorn in trying to move the program along within a group that was manufacturing a dozen other major products. The timetable continued to slip.

As GM-10 was finally ready to reach the market in 1988, Dickenson was given a new assignment, and a third program manager, Paul Schmidt, was given the task of overseeing the launch. His job was to debug the four advanced-technology assembly plants designated to build GM-10 and coordinate the vast marketing and promotional apparatus. In addition, he had to deal with the many running changes in the design of the cars. These changes were introduced after launch to increase consumer satisfaction, decrease warranty costs, and streamline factory operations.

The first GM-10 model, the Buick Regal two-door coupe, reached buyers in the spring of 1988, seven years after the initial decision to proceed with the project and two years after the original deadline. The Olds Cutlass Supreme and the Pontiac Grand Prix two-door models came next early in 1989. The last model in the range, the Buick Regal four-door sedan, finally reached dealer showrooms in the summer of 1990, nine years after the GM-10 program began. Meanwhile, Ford had launched its Taurus and Sable models as expected at the end of 1985, and Honda had gone through two generations of its Accord model, moving it up in size almost to match the physical dimensions of the GM-10 cars.

Not surprisingly, the GM-10's, although generally agreed to be competent, have encountered tough competition in the market. By 1986, GM had decided that the 1.6-million-unit annual production goal for the existing A and G bodies was not realistic and had scaled back production plans for the GM-10's to 1 million units annually (and from seven final assembly plants to four). The company wanted to reach this goal by 1990. In fact, 1989 sales were only at about 60 percent of the planned level, meaning that

# DESIGNING THE CAR

even with continuing sales of some A bodies GM lost 700,000 units of sales in its largest selling segment during the 1980s.

What's more, as we saw in the last chapter, the GM-10 cars are neither easy nor inexpensive to make. Thus what was formerly one of GM's most profitable areas—the intermediate segment—no longer pulls its weight. Indeed, the A bodies that GM-10 was to replace have proved much more profitable in the late 1980s, and the company now plans to continue the production of the Oldsmobile and Buick variants indefinitely.

## THE HONDA ACCORD: LEAN PRODUCT DEVELOPMENT

At the beginning of 1986, when the GM-10 program was already four years along, Honda began planning its own product for the intermediate segment, the fourth-generation Accord, to be launched in the fall of 1989 as a 1990 model. Since its introduction in 1976, the Accord had been the key to Honda's success in export markets and had grown steadily from subcompact to intermediate size to reflect the increasing incomes and family size of its loyal buyers.

Honda's product development process is quite different from GM's. In 1985, Tateomi Miyoshi was appointed Large Project Leader (LPL) for the new Accord and given powers far surpassing any Robert Dorn ever dreamed of. While Honda also uses a matrix, in which each project member is on loan from a functional department, Miyoshi was told to borrow appropriate people from each of the relevant departments and transfer them to the Accord project for its life. Rather than coordinating, Miyoshi's task was, clearly, to manage. He could move the project along rapidly, because all the necessary resources were under his direct control.

As the Accord plan was finalized, it became clear that the car would serve different market demands in different parts of the world. For the U.S. market, both a two-door coupe and a station wagon would be important, as would a four-door sedan. For the Japanese market a four-door hardtop would be needed along with the sedan and the coupe, the latter to be imported from the United States. Finally, Europe would be served by the sedan from Japan and the coupe and station wagon from the United States. In addition, Honda needed slightly different versions of each car for its separate Honda and Vigor distribution channels in Japan.

Honda therefore decided to subdivide its development work into one Japanese team responsible for the basic car (including the four-door sedan) and two sub-teams, one in the United States responsible for the coupe and station wagon variants, and one in Japan responsible for the four-door hardtop. The coupe and station wagon were to be produced exclusively in the United States at Honda's Marysville, Ohio, complex, using production tools designed and built at Marysville, while the sedan was to be built in both Japan and the United States, and the hardtop in Japan only.

Once the product plan was set, the Honda team steamed ahead at breakneck speed and with no interruptions. While team members continued to work closely with their functional departments, for the reasons we just mentioned, Miyoshi and practically every member of the team continued in their jobs until well after the new model was launched on schedule in the fall of 1989. They then returned to their functional departments or were assigned to a new product-development project, perhaps the next generation Accord, set for introduction in the fall of 1993.

While a conservative design, the Accord has been a resounding success in the marketplace, particularly in North America. In fact, since 1989 it has been the largest-selling model in North America, a position always held during the previous eighty years by a GM or Ford product.

## A SNAPSHOT OF PRODUCT DEVELOPMENT AROUND THE WORLD

The GM-10 and Accord cases suggest a striking difference between the lean- and mass-production approaches to product development and the consequences for competitive success. But they are only two examples, and it would be dangerous to draw firm conclusions on such limited, if provocative, evidence. Fortunately, just as we were launching our research in 1986, Professor Kim Clark at the Harvard Business School was undertaking a worldwide survey of product-development activities in the motor industry. With the help of Takahiro Fujimoto, a Ph.D. candidate at the Business School, Clark surveyed practically every auto assembler in North America, Japan, and Western Europe.[1] He asked how many hours of engineering effort were needed and how much time it had taken to produce their most recent new products. Though our projects were entirely separate in inspiration, funding, and conduct, we've benefited from lengthy discussions with the Clark team, and their work complements our global surveys of factory practice and supply-system management.

Clark and his team initially faced the same problem as we did in the assembly plant: How do you ensure an apples-to-apples comparison? Automobile development projects can vary greatly in the size and complexity of the vehicles, the number of different body styles spun off a base model (or "platform," in car talk), the number of carryover parts from previous models, and the number of shared parts from other models in a producer's range. As we have noted, carryover and shared parts require much less engineering than totally new parts. Since they are already developed, they often need only minor modifications to fit into a new model.

After they made adjustments for these variables, the Clark team's findings were simple and striking. Based on twenty-nine "clean sheet" development projects reaching the market between 1983 and 1987, Clark found that a totally new Japanese car required 1.7 million hours of engineering effort on average and took forty-six months from first design to customer deliveries.[2] (By "clean sheet," we simply mean that these were cars with totally new bodies, although some used carryover or shared engines.) By contrast, the average U.S. and European projects of comparable complexity and with the same fraction of carryover and shared parts took 3 million engineering hours and consumed sixty months. This, then, is the true magnitude of the performance difference between lean and mass production: nearly a two-to-one difference in engineering effort and a saving of one-third in development time.

Perhaps the most remarkable feature of Clark's survey is his finding that lean product-development techniques simultaneously reduce the effort and time involved in manufacturing. This fact turns on its head one of our most common assumptions, one that is based on seventy years' experience with mass production: namely, that any project can be speeded up in a crisis but that doing so greatly increases the cost and amount of effort.

We're all familiar with the refrain: "Sure I can get it ready sooner, but it's going to cost you a fortune!" We suggest that "faster is dearer" will now join "quality costs more" (an idea debunked in Chapter 4) on the junk heap of ideas left over from the age of mass production.

## THE TECHNIQUES OF LEAN DESIGN

It's wonderful to know that new products can now be produced faster with less effort and fewer errors. However, as we've noted, innovations are most useful when they are available to everyone, and, as we have seen, practices at General Motors and the other mass-producers lag far behind. What then are the precise techniques of lean design that the best auto firms use and how can they be transferred to existing mass-producers?

To find the answer we began by looking at Clark's and Fujimoto's work, but then asked Antony Sheriff and Kentaro Nobeoka of our IMVP staff to conduct additional investigations. Sheriff, a former product planner with Chrysler, had become disillusioned with American approaches to product development when he joined the IMVP, while Nobeoka was a product planner on leave from Mazda to pursue a Ph.D. at MIT. Between the two of them they had a vast knowledge of product development as it looks from the inside.[3]

From Clark's and Fujimoto's work and our own investigations, we conclude that there are four basic differences in design methods employed by mass and lean producers. These are differences in leadership, teamwork, communication, and simultaneous development. Taken together, lean techniques in these four areas make it possible to do a better job faster with less effort.

### Leadership

First, let's look at the leadership of a project. The lean producers invariably employ some variant of the *shusa* system pioneered by Toyota (termed the "large-project leader," or LPL, system at Honda). The *shusa* is simply the boss, the leader of the team whose job it is to design and engineer a new product and get it fully into production. In the best Japanese companies the position of *shusa* carries great power and is, perhaps, the most coveted in the company. True, employees may seek the position as a stepping-stone to the top. However, for those who truly love to make things, the job brings extraordinary satisfaction. In fact, it's the best position in the modern world from which to orchestrate all

# DESIGNING THE CAR

the skills needed to make a wonderfully complex manufactured product, such as the automobile, come into being.

One might even say that the *shusa* is the new supercraftsman, directing a process that now requires far too many skills for any one person to master. Oddly, while we're used to thinking of dedicated teamwork as the ultimate sublimation of individuality, new products inside the Japanese auto industry are commonly known by the *shusa*'s name: "That is Fuji-san's car" or "Akoika-san has really stamped his personality on that car" are phrases commonly heard within Japanese companies. Perhaps, after all, we cannot escape a human need for craftsmen to exist. However, in an era when the skills involved are not so much technical as social and organizational—and far beyond the grasp of any individual—craftsmen must now take the form of the *shusa*.

Western mass-producers also have development team leaders, as we saw in the GM-10 example. What's the difference between the two systems? We believe it lies in the power and career path of the team leader. In Western teams, the leader is more properly called a coordinator, whose job it is to convince team members to cooperate. It's a frustrating role, because the leader really has limited authority, so few team leaders report enjoying it. Indeed, many company executives view the job as a dead end in which success leads to little reward and failure is highly visible (as we saw in the story about the GM-10 project, which opens this chapter).

What's more, the team leader is in an extremely weak position to champion a project within the company. It's common in Detroit, Wolfsburg, and Paris for top management to override the team leader about the specifications and feel of the product—often repeatedly during the course of development. That this happens is understandable, given senior management's role of juggling other corporate needs as market conditions change. However, in the worst case—and all too frequently, particularly in the United States—the result is a product with no personality or distinction that the company must sell solely on the basis of low price.

## Teamwork

The problem here becomes clearer by looking at the second element of lean design, the tightly knit team. As we saw, in the lean-development process, the *shusa* assembles a small team, which is then assigned to a development project for its life. These employees come from functional departments of the company—market assessment, product planning, styling, advanced engineering, detail engineering (body, engine, transmission, electrical), production engineering, and factory operations. They retain ties to their functional department—it's vital that they do so, as we explained earlier in this chaper—but for the life of the program they are clearly under the control of the *shusa*. How they perform in the team, as judged by the *shusa*, will control their next assignment, which will probably be another development team.

By contrast, in most Western companies a development project consists of individuals, including the team leader, who are on short-term loan from a functional department. Moreover, the project itself is moved from department to department along a sort of production line, which leads from one end of the company to the other. That is, the project is actually picked up and moved from the marketing department to the engineering divisions, and then to the factory operations department during its life, in the same way that a car moves from the welding to the painting to the assembly department in the assembly plant. So it is worked on by totally different people in each area.

The members of the team know that their career success depends on moving up through their functional specialty—getting promoted from chief piston engineer to deputy chief engine engineer to chief engine engineer, for example—and they work very hard in the team to advance the interest of their department. In other words, being a member of the GM-10 team, say, doesn't lead anywhere. The team leader will never see an employee's personnel records, and the leader's performance evaluation won't make much difference to the employee's career. Key evaluations will come from the head of the employee's functional division, who wants to know, "What did you do for my department?" As a result, discussing the best way to achieve harmony between the engine and the body, for example, easily can disintegrate into a politicized debate between the interests of the engine engineering department and the body engineering department.

The continuity in the Japanese development teams is reflected in another of Clark's and Fujimoto's findings. They discovered that about 900 engineers are involved in a typical project in an American or European company over its life, while a typical Japanese team enlists only about 485.[4] What was more, those

Japanese firms most committed to the *shura* system (which Clark and Fujimoto term "heavyweight" team management) needed an average of only 333 team members, while the Western firms with the weakest teams (mostly in Germany) needed an average of 1,421 staff members during the life of a project. The Japanese use fewer people partly because efficient organization requires fewer bodies, but also because there is so little turnover in the Japanese teams. Because Western department managers view team members as simply the representatives of their home department in the development process, they show little concern about frequently recalling staff as other needs for their skills crop up in their own department. For the team, however, these recalls mean a great loss because much of the essential knowledge of a development team lies in the shared viewpoints and experiences of team members over an extended period.

**Communication**

This point brings us to the third feature of lean design—communication. Clark and Fujimoto found that many Western development efforts fail to resolve critical design trade-offs until very late in the project. One reason is that U.S. team members show great reluctance to confront conflicts directly. They make vague commitments to a set of design decisions—agreeing, that is, to try to do something as long as no reason crops up not to. In Japan, by contrast, team members sign formal pledges to do exactly what everyone has agreed upon as a group. So conflicts about resources and priorities occur at the beginning rather than at the end of the process. Another reason is that a design process that is sequential, going from one department to the next rather than being kept at team headquarters, makes communication to solve problems very difficult in any case.

The result is a striking difference in the timing of the effort devoted to a project. In the best Japanese lean projects, the numbers of people involved are highest at the very outset. All the relevant specialties are present, and the *shusa*'s job is to force the group to confront all the difficult trade-offs they'll have to make to agree on the project. As development proceeds, the number of people involved drops as some specialties, such as market assessment and product planning, are no longer needed.

By contrast, in many mass-production design exercises, the number of people involved is very small at the outset but grows to a peak very close to the time of launch, as hundreds or even thousands of extra bodies are brought in to resolve problems that should have been cleared up in the beginning. The process is very similar to what we saw in the assembly plant: The mass-producer keeps the line moving at all costs but ends up doing massive amounts of rework at the end, while the lean producer spends more effort up front correcting problems before they multiply and ends up with much less total effort and higher quality in the end.

### Simultaneous Development

The final technique separating lean from mass production in product development is simultaneous development. To see what we mean by this term, let's take the example of die development.[5]

As we pointed out in Chapter 4, practically every car and light truck built in the world today has a body constructed of stamped steel panels. The heavy metal forms, called dies, needed to press finished body panels out of sheet steel are among the most complex and expensive tools in the industrial world. They are made of exotic steel alloys for extreme strength and hardness and must be formed to tolerances of microns across continuously curving surfaces. What's more, the matching faces of the die (the upper and lower or "male" and "female" elements) must mesh with absolute precision. Otherwise, the sheet steel will tear or even melt to the face of the dies as the two pieces come together under tons of pressure.

The mass-production approach to die-making has been simple: Wait until the product designers give precise specifications for the stamped part. Then, order an appropriate block of steel in the die-production department and cut it, using expensive, computer-driven, die-cutting machines. Because cutting proceeds through many steps involving many machines, this process means that the dies pile up waiting for the next machine to become available. Total development time, from the first day that product designers order a new set of dies until the dies begin stamping panels for production cars, is about two years.

By contrast, the best lean producers—and they're all Japanese but no longer only in Japan (Honda is designing and cutting dies

DESIGNING THE CAR                                                            117

for its Marysville, Ohio, plant at Marysville)—begin die production at the same time they start body design. How can they? Because the die designers and the body designers are in direct, face-to-face contact and probably have worked together in previous product-development teams.

The die designers know the approximate size of the new car and the approximate number of panels so they go ahead and order blocks of die steel. Then they begin to make rough cuts in the steel, so it's ready to move to final cutting as soon as the final panel designs are released.

This process, of course, involves a considerable degree of anticipation. The die designer must understand the panel-design process as well as the panel designer does and be able to anticipate accurately the panel designer's final solution. When the die designer is correct, development time is drastically shortened. When the die designer is wrong (an infrequent occurrence), the company pays a cost penalty. Still, the original schedule can be met by giving the catch-up die priority routing through the cutting process.

Also, the lean die makers seem to be much better at scheduling production in the die cutting shop. Their solution should come as no surprise if you remember the example of Ohno's stamping shop from Chapter 3: The die cutters have special, quick-change cutting tools, allowing one machine to handle many different types of cuts, so the dies that are being cut spend much less time in queues.

What's the end result of this intense communication between panel designers and die makers plus accurate anticipation by the die makers and clever scheduling of flexible cutting machines? It means that the best lean producers in Japan (and in Ohio) can produce a complete set of production-ready dies for a new car in one year, exactly half the time needed in typical mass-production die-making.[6] Not surprisingly, this process requires fewer tools, lower inventories (since the key element, the expensive die steel, is in the shop only half as long), and less human effort.

### PRODUCTION-DEVELOPMENT BOX SCORE: LEAN VERSUS MASS

Figure 5.1 sums up all the advantages of lean product development in the form of a box score.

#### FIGURE 5.1

**Product Development Performance by Regional Auto Industries, Mid-1980s**

|  | Japanese Producers | American Producers | European Volume Producers | European Specialist Producers |
|---|---|---|---|---|
| Average Engineering Hours per New Car (millions) | 1.7 | 3.1 | 2.9 | 3.1 |
| Average Development Time per New Car (in months) | 46.2 | 60.4 | 57.3 | 59.9 |
| Number of Employees in Project Team | 485 | 903 | 904 | |
| Number of Body Types per New Car | 2.3 | 1.7 | 2.7 | 1.3 |
| Average Ratio of Shared Parts | 18% | 38% | 28% | 30% |
| Supplier Share of Engineering | 51% | 14% | 37% | 32% |
| Engineering Change Costs as Share of Total Die Cost | 10–20% | 30–50% | 10–30% | |
| Ratio of Delayed Products | 1 in 6 | 1 in 2 | 1 in 3 | |
| Die Development Time (months) | 13.8 | 25.0 | 28.0 | |
| Prototype Lead Time (months) | 6.2 | 12.4 | 10.9 | |
| Time from Production Start to First Sale (months) | 1 | 4 | 2 | |
| Return to Normal Productivity After New Model (months) | 4 | 5 | 12 | |
| Return to Normal Quality After New Model (months) | 1.4 | 11 | 12 | |

*Source:* Kim B. Clark, Takahiro Fujimoto, and W. Bruce Chew, "Product Development in the World Auto Industry," *Brookings Papers on Economic Activity*, No. 3, 1987; and Takahiro Fujimoto, "Organizations for Effective Product Development: The Case of the Global Motor Industry," Ph.D. Thesis, Harvard Business School, 1989, Tables 7.1, 7.4, and 7.8

When we review this box score, we can see several additional advantages of lean design. For one, lean design results in a much higher fraction of projects coming into production on time. Indeed, five of six Japanese projects reach the market on the timetable laid out at the beginning of development, while only half of American projects come in on time. The GM-10 project was worse than average in its slippage, but hardly unusual.

Another advantage lies in the ability of the lean factory to absorb new products without paying a productivity penalty. Many Western analysts have been badly misled by the slow start-up schedules ("ramp-up rates," in car talk) of the Japanese transplants in North America and Europe. What they fail to realize is that these facilities are building up the social process of produc-

# DESIGNING THE CAR

tion step by step and this takes time. For example, Toyota executives at the Georgetown, Kentucky, plant say that it will take a decade for the plant to fully master the Toyota Production System. To ensure than no one takes short cuts, they've built up the plant's production rate very slowly, stopping as necessary to get each step right rather than rushing ahead and going back later to rework errors not just in the cars but in the entire production organization.

However, once lean production is fully in place in the factory, it's easy to introduce new products developed by a lean-development process. For example, Japanese plants taking on new models regain their previous productivity level in four months, while the Americans need five, and the Europeans need an entire year.[7]

Much more striking is the difference in quality. Japanese lean-production plants can introduce new lean designs with only a small blip in delivered quality, while American and European plants struggle for a year to get quality back to its original level, which is lower than that of the Japanese to begin with.

## THE CONSEQUENCES OF LEAN DESIGN IN THE MARKETPLACE

What do companies that have mastered lean design do to take advantage of their strength in the marketplace? Obviously, they will offer a wider variety of products and replace them more frequently than mass-production competitors. And this is exactly what has been happening in the auto industry across the world in the 1980s.

In Figure 5.2 we summarize the number of models the Japanese car companies were selling worldwide between 1982 and 1990.[8] We then compare this number with the total worldwide product offerings by the U.S.-headquartered producers and the five high-volume European car companies (PSA, Renault, Fiat, Rover, and Volkswagen). We provide a separate calculation for five smaller European specialist firms: BMW, Mercedes, Volvo, Saab, and Jaguar.

The trend is striking. The Japanese firms are using their advantage in lean production to expand their product range rapidly, even as they renew existing products every four years. Between 1982 and 1990 they nearly doubled their product portfolio from forty-seven to eighty-four models.

**FIGURE 5.2**

**Number of Models and Average Model Age by Regional Origin of Producers, 1982–1990**

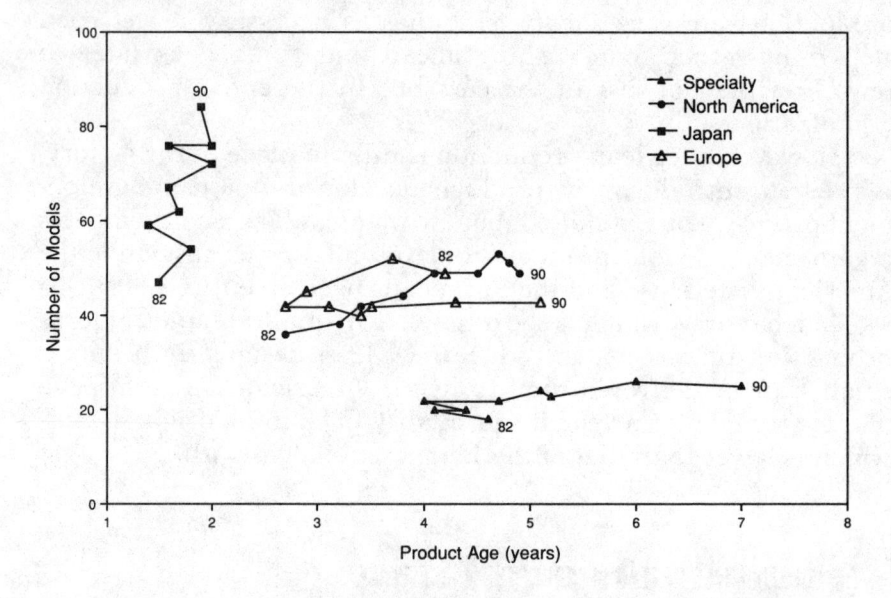

*Note:* Companies are grouped into categories based on the location of their headquarters. All products developed by each company within the three major regions are included in the count for the headquarters region. Thus the cars developed by General Motors and Ford in Europe are included in the "American" count. Models developed outside the three major regions, with the exception of the Ford Capri from Australia, are excluded. Thus the models developed by General Motors, Fiat, Ford, and Volkswagen in Brazil, and the models developed by Ford and GM Holden's in Australia are not counted.

The model count includes all automobiles and car-derived, front-wheel-drive mini-vans. It excludes rear-wheel-drive mini-vans, sport/utility vehicles, and trucks.

A "model" is defined as a vehicle with entirely different external sheet metal from any other product offered by a company. Thus GM-10 is counted as four models and Ford Taurus/Sable is counted as two models. Two-, three-, four-, and five-door variants and station wagon versions of the same car are counted as one model.

Average product age has been weighted by sales volume because a number of very low volume products in Europe and Japan are continued in production for very long periods. Products from craft producers, such as Ferrari and Aston Martin, and models in production for more than twenty years, such as the Morris Mini and Citroën Deux Cheveux, have been excluded.

*Source:* Calculated by Antony Sheriff from product data in *Automobile Review,* Geneva, 1990 and previous years

At the same time, the European volume firms were pursuing old-fashioned mass-production strategies as they struggled to absorb the companies taken over in the 1970s and 1980s. They reduced their models on offer slightly, from forty-nine to forty-three, and allowed remaining models to age markedly. Specifically, PSA (Peugeot) rationalized the product offerings of Citroën and Chrysler Europe while Fiat consolidated the offerings of Alfa Romeo. Recently Volkswagen has absorbed the Spanish producer Seat (which previously built Fiat designs under license), Volvo and Renault have agreed to collaborate on auto-making activities, and General Motors has become the senior partner in a joint venture involving Saab's automotive operations. These events suggests that one more round of product rationalization may be carried through in Europe in the early 1990s.

The Americans by contrast have managed a substantial increase in their product ranges, from thirty-six to fifty-three models, but at a cost, as you can see along the bottom axis of Figure 5.2. The numbers here refer to the average number of years the average product has been in production. In the case of the Japanese producers, this number is between 1.5 and 2.0 years—about what one would expect from companies with policies of replacing every model every four years. For the Americans, by contrast, the average age has risen from 2.7 to 4.7 years, suggesting that the average model is now kept in production for nearly ten years rather than the eight years common in the past. The reason, we believe, is simply that the Americans, with their inefficient product-development processes, are finding they do not have the money or engineers to expand their product range and renew their products frequently.

A quick look at the North American motor-vehicle market, as shown in Figure 5.3, indicates that the Japanese strategy of the 1980s is likely to continue in the 1990s. As of the 1991 model year, the Japanese companies still offer no products in the full-size car, van, and pickup truck classes. Similarly, their range of offerings in the European luxury and sport specialty classes is still very modest, despite the recent excitement about Lexus, Infiniti, and Acura. Large cars and pickup trucks are the most profitable areas of the entire world market. So it would be remarkable if the Japanese producers did not move ahead rapidly to complete their product offerings in the larger size classes of cars, trucks, and vans, and perhaps to develop new market segments as well.

At the same time, the European volume producers will soon

## FIGURE 5.3

**Producer Shares of the American Motor Vehicle Market, 1989**

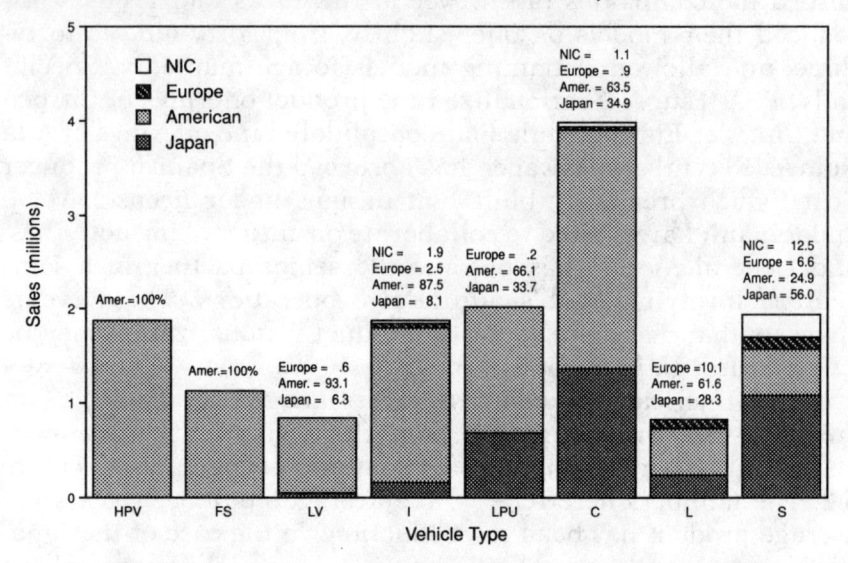

*Note:* These are market shares and sales volumes for producers headquartered in Japan, Europe, and North America and for producers headquartered in the newly developing countries, principally Korea. All vehicles sold by those producers are included in their share, wherever the vehicles were produced. Thus the Japanese share includes vehicles assembled at their North American plants.

HPV = full-size pickup trucks and full-size vans
FS   = full-size cars priced below $25,000 U.S.
LV   = mini-vans
I    = intermediate-size cars
LPU  = small pickup trucks
C    = compact cars
L    = luxury cars of any size priced above $25,000 U.S.
S    = subcompact cars

*Source:* Calculated by the authors from *Ward's Automotive Reports*

complete their product consolidation process, and companies from all three regions are likely to be adding products such as mini-mini vans in the smaller size classes. The consequence? All surviving producers will offer a wider product range in the 1990s, but unless the Western mass-producers transform their product-development systems, the Japanese producers will be able to widen their range much more rapidly and, even as they do so, be able to keep their existing products fresh by replacing them every four years.

# DESIGNING THE CAR 123

This trend has striking implications for production volumes, both annually for each model and cumulatively over each model's life. Figure 5.4 shows the average annual production volume for all models produced worldwide by the companies headquartered in each region. American volume per model has been falling, not only because the number of models offered has increased but because the Americans as a group have been losing market share and total volume. Nevertheless, the Americans still produce 60 percent more units of their average model each year than do the Japanese. European production per model (among the volume producers) has also been rising, partly due to model consolidation and partly due to the extremely strong European car market. The Europeans are currently also producing 60 percent more copies of the average model each year than the Japanese.

Figure 5.5 takes the analysis one step further. We have doubled average product age (shown in Figure 5.2), then multiplied by average annual production volume per model (shown in Figure

**FIGURE 5.4**

**Annual Production Volume of the Average Model, by Regional Producers, 1982–1990**

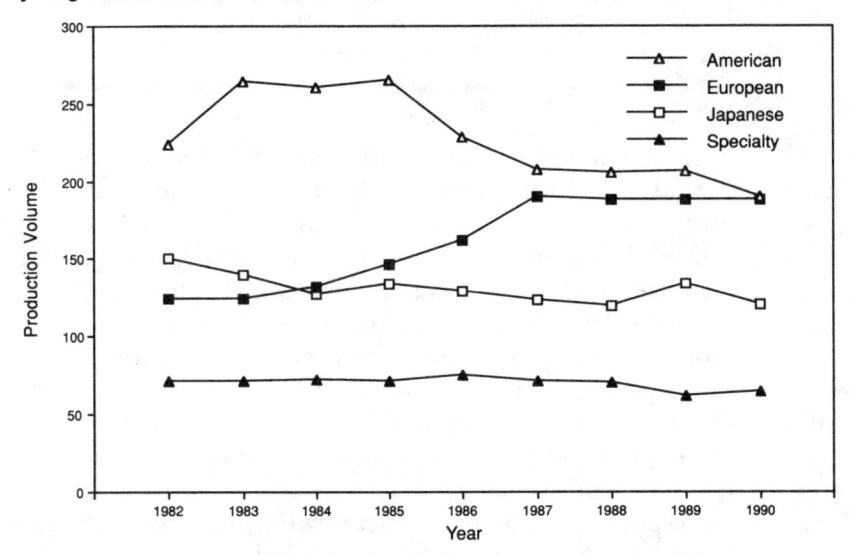

Note: "Models" are as defined in Figure 5.2. All production of a given model worldwide is summed under the headquarters region. Production for 1990 has been estimated.

Source: Calculated by Antony Sheriff from production data from PRS

**FIGURE 5.5**

**Estimated Life-of-the-Model Production Volume, by Region, 1982–1990**

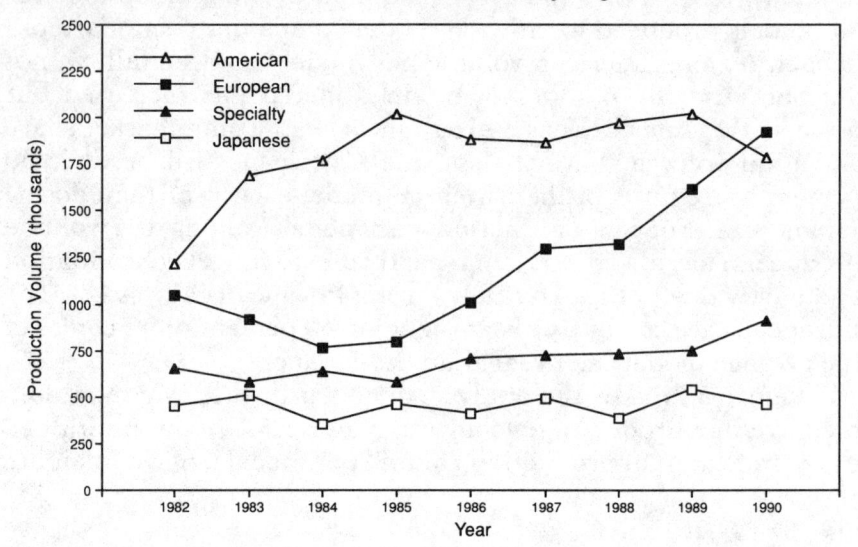

*Note:* "Models" are as defined in figures 5.2 and 5.4. This figure has been estimated by doubling the average product age shown in Figure 5.2 and multiplying by annual production volumes shown in Figure 5.4. Some type of estimation is unavoidable, because most of the models counted in the figure will continue in production for a number of years.

*Source:* Calculated by Antony Sheriff from PRS and *Automobile Review* data

5.4) for the Americans, the European volume producers, the European specialists, and the Japanese. Because the Japanese models—with very few exceptions—remain in production only four years compared with eight to ten years for the American and European volume producers, it is not surprising that the Japanese are producing only one quarter as many copies of each car during its production life. What is surprising is that the very long model lives of the European specialist products result in 50 percent more copies being produced during their lives than of the average Japanese "mass market" car. Who are the real "specialists" in today's world auto industry?

A final perspective on lean product strategy emerges when we shift our focus from production volumes worldwide to what has been happening in a specific market, notably the U.S. car, van, and truck market. We have already seen in Figure 5.3 that a broad array of market segments are now served with a surprisingly even level of sales across segments. Figure 5.6 shows the dramatic

# DESIGNING THE CAR

increase in the number of products on offer since the heyday of mass production in 1955 and the continually declining number of sales per product. (The years 1955, 1973, and 1986 have been used because these were years of peak demand in this highly cyclical market. To make the 1989 sales-per-product figures comparable to the earlier years, we have assumed that 1989 sales were at the 1986 level when they were actually about 9 percent lower.)

**FIGURE 5.6**

**Fragmentation of the American Auto, Van, and Light Truck Market, 1955–1989**

|  | 1955 | 1973 | 1986 | 1989(2) |
|---|---|---|---|---|
| *Total:* | | | | |
| Products on Sale (1) | 30 | 84 | 117 | 142 |
| Sales/Product (000s) | 259 | 169 | 136 | 112 |
| Share of Market Captured by 6 Largest-Selling Products | 73 | 43 | 25 | 24 |
| *American Products: (3)* | | | | |
| Number on Sale | 25 | 38 | 47 | 50 |
| Sales/Product (000s) | 309 | 322 | 238 | 219 |
| *European Products: (3)* | | | | |
| Number on Sale | 5 | 27 | 27 | 30 |
| Sales/Product (000s) | 11 | 35 | 26 | 18 |
| *Japanese Products: (3)* | | | | |
| Number on Sale | 0 | 19 | 41 | 58 |
| Sales/Product (000s) | 0 | 55 | 94 | 73 |

*Notes:*
(1) A *product* is defined as a vehicle selling more than 1,000 units in the U.S. market annually that shares no external panels with and has a different wheelbase from any other vehicle in a producer's range. Thus Ford Taurus and Mercury Sable are counted as one product, as are the four GM-10 cars sold under the Chevrolet, Pontiac, Oldsmobile, and Buick badges. Although the Ford and GM cars have entirely different exterior sheet metal, they share the same wheelbase and a large number of under-the-skin structural items along with many mechanical components. Note that the *products* counted here are not the same as the *models* defined and counted in figures 5.2, 5.4 and 5.5.
(2) 1955, 1973, and 1986 were chosen because they were years of peak volume in the highly cyclical American motor-vehicle market. Sales in 1989 were down 9 percent compared with 1986, so using 1989 volume would not provide a good comparison of the trend in average sales per product. We have therefore used 1986 sales volume to calculate the total sales per product in 1989. Also, we have taken 1989 market shares by regional producers, multiplied these by 1986 overall sales to gain sales by regional producer, and then used these figures to calculate average sales per product by region for 1989.
(3) These are products built anywhere in the world for sale in the U.S. market by firms headquartered in each region. Thus the Volkswagen Fox model built in Brazil is counted as European, and the Ford Merkur models built in Germany are counted as American.

*Source:* Calculated by the authors from sales data in *Ward's Automotive Reports*, January 8, 1990 (for 1989) and in the *Ward's Automotive Yearbook* (for 1955, 1973, and 1986).

In fact, the ultimate destination of lean production in terms of the variety of product offerings is unknown. We have recently talked with Japanese auto executives planning continuing and major reductions in their target volume for sales of each product. In the extreme case, although not for several decades, is it possible that we will return full circle to the world of craft production, where each buyer was able to custom order a vehicle suited to his or her precise needs?

This possibility is graphically portrayed in Figure 5.7, which shows that at the beginning of the automotive age there was an extraordinary range of products for sale, with production and sales volume for the average car very low. Often, as we saw at Panhard et Levassor, each car was a one-of-a-kind custom-built to the wishes of its owner. Under Henry Ford the volume per car zoomed, to 2 million copies per year for the Model T, yet the exit of practically all craft producers from the industry meant that product variety fell from thousands to dozens of offerings. Classic mass production, under Alfred Sloan, increased product variety modestly, but the world awaited the arrival of lean production for a true renaissance in consumer choice. And the end is not yet in sight.

**FIGURE 5.7**

**The Progression of Product Variety and Production Volume in the Auto Industry**

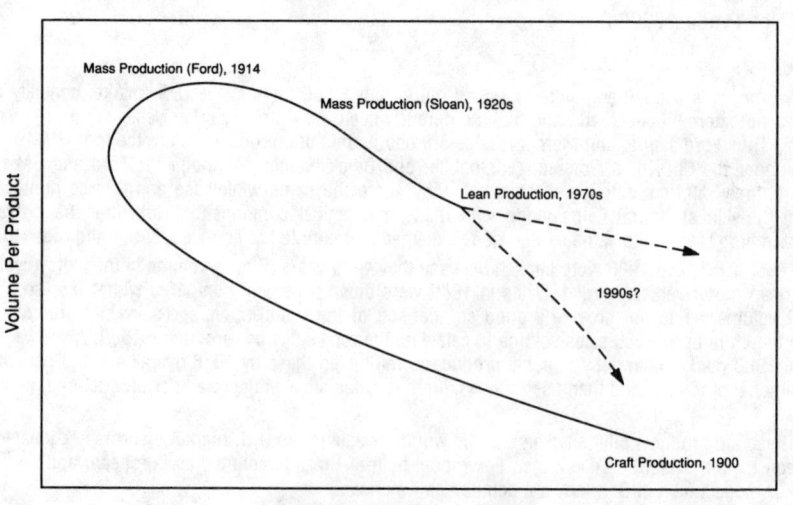

## WHERE LEAN DESIGN GOES NEXT

When we present these findings to Western car companies, senior executives often say that the Japanese focus on short model cycles and wider product variety is interesting, but only as a curiosity, not a threat. "They will never be able to sustain this pace, and consumers will soon grow tired of short cycles and too many choices," said one senior European executive (a position that ignores the success of Japanese firms using similar strategies of product proliferation with a host of other consumer products—motorcycles, cameras, watches, consumer electronics).

The arrival on the scene of Japanese luxury cars seems even to have strengthened this position. As another executive remarked: "Buyers of luxury cars don't want constant model changes, since it undermines the resale value. The Japanese will have to stop this."

We respectfully disagree. We think of lean product development as a multifaceted capability, which has fundamentally changed the logic of competition in this industry. The producers in full command of these techniques can use the same development budget to offer a wider range of products or shorter model cycles—or they can spend the money they save by implementing an efficient development process for developing new technologies. If luxury car buyers resist a short model cycle, the lean producer can concentrate on a wider variety of models. If wider model variety doesn't attract customers, new technologies may—perhaps electronic suspension, a body so immune to rust that it carries a lifetime guarantee, or even a new type of engine. And in every case, the shorter development cycle will make lean companies more responsive to sudden changes in consumer demand. The choice, and the advantage, will always lie with the lean producer. This becomes more apparent when we contrast lean- and mass-production approaches to the development of new technologies.

## INVENTING SOMETHING NEW

The people involved in the product-development efforts we've just looked at were engaged in routine problem-solving. They combined existing components and proven engineering principles to develop new products that were up-to-date and in tune with current consumer desires. In other words, they solved problems without having to engineer anything that was fundamentally new.

But what happens when the old solutions no longer work—when the external world changes in such a way that existing components and design principles are no longer adequate to do the job? And, what does a company do when competition stiffens and it needs something better than a me-too solution to maintain its market position?

This is the job for research as opposed to development—the conscious process of inventing, perfecting, and introducing something that is new. As we will see, lean producers approach the problem in a very different way from mass-producers.

### Invention in Mass Production

Alfred Sloan was an MIT-trained electrical engineer, so his advice on technological innovation may surprise us. In his memoirs, *My Years with General Motors*, he had this to say on the topic: ". . . it was not necessary to lead in technical design or run the risk of untried experiments [provided that] our cars were at least equal in design to the best of our competitors in a grade."[9]

When he wrote these words, in retirement in the early 1960s, Sloan had come to perceive a special problem with innovation because of the size and market dominance of GM. By that time, when GM had sewn up half the North American auto market, any truly epochal innovation—say, a turbine-powered truck or a car with a plastic body—could have bankrupted Ford and Chrysler. The auto makers' plight would certainly have attracted the attention of a U.S. government intent on preventing a monopoly in its largest industry. So caution made sense. GM hardly wanted to innovate its way to corporate dismemberment.[10]

However, the way GM and the other large mass-producers—including those in Europe—went about organizing their fundamental research made it highly unlikely that they would come up with startling innovations in any case. Unfortunately, they have learned this sad fact only very recently.

In thinking about innovation, Sloan carried Henry Ford's ideas on the division of labor to their logical extreme. He decided to concentrate the scientists and engineers who were working on advanced preproduction ideas at GM's technical center outside Detroit. There, he felt, they would be free of the daily distractions of commerce and able to focus on the company's long-term needs.

Over the decades, GM built an enormous, very high-quality staff and made a number of fundamental discoveries. In the mid-1970s, in fact, GM's technological resources proved vital to the welfare of the entire world auto industry when its scientists and engineers—on very short notice—perfected the exhaust catalyst technology now used by every car company in the world to produce automobiles that meet emission standards. GM proved that when the external environment demanded quick action, it could, and can, innovate.

Unfortunately, in the absence of a crisis—a situation in which the future of the company was at stake and normal organizational barriers to the flow of information were suspended—new ideas percolated from the research center to the market very slowly. And when a crisis did occur, the lack of day-to-day contact between the thinkers in the research center and the implementers in product development often meant embarrassing gaffes. In GM's case these included the Corvair project in the late 1950s, the Vega project in the late 1960s, the X-car project in the late 1970s, and the high-tech factories for the GM-10 products in the late 1980s. In each case, innovative ideas for new products and factories foundered when implementation could not live up to the original technical targets. These results are in startling contrast with what has happened in the last decade in the lean-production companies.

## Invention in Lean Production

University-trained mechanical, electrical, and materials engineers start their careers in an interesting way at many of the Japanese lean producers.[11] They assemble cars. At Honda, for example, all entry-level engineers spend their first three months in the company working on the assembly line. They're then rotated to the marketing department for the next three months. They spend the next year rotating through the engineering departments—drive train, body, chassis, and process machinery. Finally, after they have been exposed to the entire range of activities involved in designing and making a car, they are ready for an assignment to an engineering specialty, perhaps in the engine department.

At first, they are likely to be assigned to a new-product development team. There they will do very routine work, largely adapting established designs to the precise needs of the new model. This task, as we saw in the preceding chapter, continues for up to four years.

After successfully working on a new development project, the young engineer is likely to be transferred back to the engine department to do more fundamental work, perhaps on the design of a new engine, such as the V6 and V8 units recently introduced by the Japanese producers and intended for use in a whole range of new models. (An engine-development program, like a new-model development program, requires three to four years between initial concept and actual production.)

Once the engineers successfully finish their stint on this second type of development team, some of the most promising are selected for additional academic training and then are set to work on longer-term and more advanced projects. For example, the engineer might study how to incorporate fiber reinforcements into highly stressed metal parts, such as the rods connecting the crankshaft to the pistons. In working on these projects, the engineers consult closely with academic experts on retainer to the company.

However, even these longer-term development projects have a very specific objective—to remedy some weakness in the company's products identified by the product or major-component development teams. So they are tied tightly to the needs and timetable of specific development projects. And, the work is conducted by engineers who thoroughly understand the practicalities of product development and production. To make sure that the engineers maintain their sensitivity, Honda, for example, assigns even its most advanced engineers to spend a month of each year working in one of the other functional areas of the company—the selling divisions, factory operations, supply coordination, and so forth.

The Japanese lean producers exercise extreme care not to isolate their advanced technologies from the day-to-day workings of the company and the incessant demands of the market. Based on their observations of U.S. and European mass-producers, they long ago concluded that, to be effective, engineering, even of the most advanced sort, must be tied into the key market-driven activities of the company.

## LEAN INNOVATION IN PRACTICE: LOW-TECH WEAKLINGS TO HIGH-TECH WONDERS

Just how this process can work is nicely illustrated by the evolution of Japanese engine designs during the 1980s. At the beginning of the decade, the Japanese companies faced a common problem. They had assumed that energy prices would continue to climb and that consumers would want smaller cars, so they had invested billions of dollars in the late 1970s in new engine plants for small four-cylinder engines. Instead, fuel prices fell and consumers were looking for larger cars with more power.

What to do? Engine sizes can be increased slightly using existing production tools by widening the bore of the cylinders and increasing their stroke. However, to go further—by adding cylinders or changing the engine's configuration, from, say, four cylinders in a line to six in a V arrangement—would be staggeringly expensive, because it would require junking most of the existing production tools. New billion-dollar engine plants would, in turn, drain resources from the product-development teams straining to rapidly increase the range of Japanese products. Surely, the lean producers thought, there is a quicker and easier solution.

In fact, there was. The product-development teams turned to the advanced engineering groups, which suggested introducing every available technical feature to boost the performance of the basic four-cylinder engines. These features were conceptually simple—fuel injection rather than carburetors, four valves per cylinder rather than two (to get more fuel in and more exhaust out on each stroke), balance shafts in the bottom of the engine (to damp the inherent roughness of four-cylinder designs), turbochargers and superchargers (to get more power from the same size engine), a second set of overhead cams (to make valve timing more precise), and even an additional set of cams for use at higher engine speeds (to get full power from the engine across a wide range of operating conditions).

In addition, the engineers worked very hard on what is known in the industry as refinement—paying attention to the smallest details of an engine design so that the finished engine runs smoothly and without complaint at all speeds and in all driving conditions, imitating the performance of a much larger engine.

Finally, the engineers paid endless attention to manufacturability. Because they were going against one good engineering practice—by adding parts and complexity to an already complex device—they had to work extra hard at manufacturability so the complex engines would work properly every time and entail only the minimum of extra production expense.

As these features were added during the decade of the 1980s, they had an interesting effect on public perceptions—one that was perhaps unanticipated. Even as they raised the power of the same basic engine, in some cases by a factor of two, these innovations convinced buyers, particularly in North America, that Japanese cars were now "high-tech," that they now had the most advanced features. They had grown from "low-tech" weaklings in 1980 to "high-tech" wonders by 1990 while preserving their manufacturers' basic investment in production facilities for small engines.

This perception on the part of consumers was enormously frustrating to engineers in many of the mass-production companies who knew that all these "innovations" had been around the motor industry for decades. For example, four valves per cylinder and double overhead camshafts were available on the 1924 Bentley, and superchargers were common on the largest European luxury cars in the 1930s. However, they had often been vetoed by management as too expensive or complex for production use or restricted to use in a limited range of specialty models.

What's more, when the mass-producers, particularly in the United States, tried to copy these "innovations" on a wide scale, the weaknesses of their engineering systems were exposed. In many cases it took years to introduce a comparable feature, and often it was accompanied by drivability problems or high production costs. GM, for example, lagged Toyota by four years in introducing many of the features we just listed in its Quad Four engine; it needed two more years to reach a high level of refinement. Even then, manufacturing bottlenecks meant that this engine was available in only a narrow range of GM cars using four-cylinder engines.

### LEAN PRODUCTION VERSUS MASS IN RESEARCH AND DEVELOPMENT: SOME NUMERICAL COMPARISONS

Given the contrasting approaches to innovation, it's not surprising that the examples we cited are typical and that performance in the development of new technologies differs systematically. In

*DESIGNING THE CAR*

particular, it is not surprising that the U.S. companies manage to spend more on their research activities, as shown in Figure 5.8, but badly trail the Japanese companies in a key indicator of technological strength—patenting—as shown in Figure 5.9. (The patent data are for all patents registered in the United States by motor-vehicle and supplier firms from the different regions of the world.)[12]

What's more, in the last decade the Japanese lean producers have consistently outpaced the Americans and even the Europeans in bringing these patented innovations to market.

**FIGURE 5.8**

**Annual Spending on Motor Vehicle Research and Development, by Region, 1967–1988**

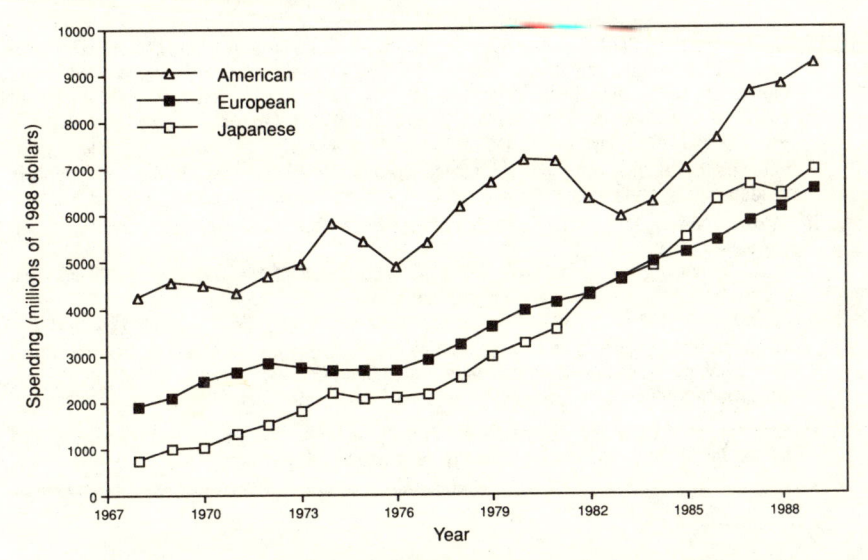

*Note:* The figures are for worldwide spending on research and development by firms in the motor vehicle industry grouped by headquarters region. Thus General Motors' worldwide spending is consolidated under "American" and Volkswagen's worldwide spending is grouped under "European."

Figures are constant 1988 dollars at 1988 exchange rates.

*Source:* Calculated by Daniel Jones from the Organization for Economic Cooperation and Development's annual "Compilation of Surveys of R&D by Member Governments"

#### FIGURE 5.9

## Motor Vehicle Industry Patenting, 1969–1986

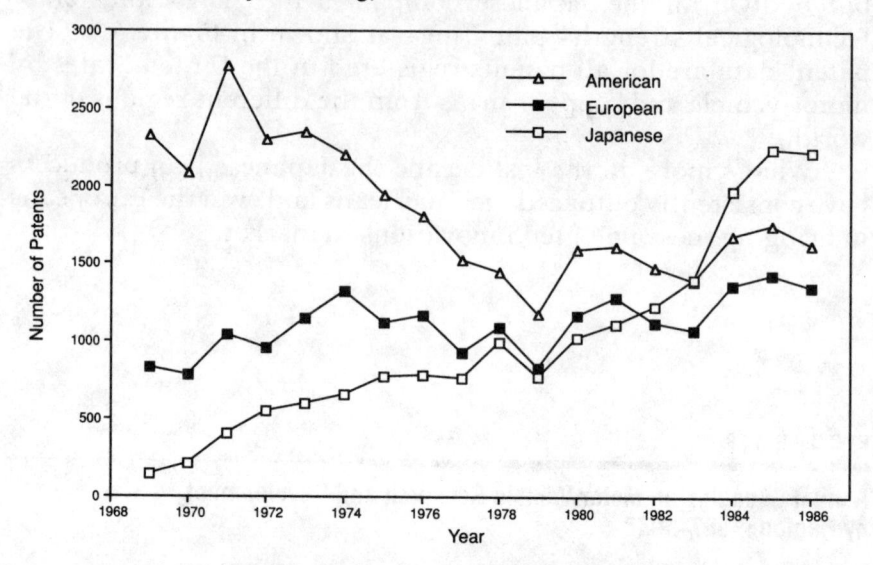

*Note:* Figures are for patents granted by the U.S. Patent Office to assembler and supplier firms located in each main region. In case of subsidiaries whose parent is headquartered in one region but which operate in another region, the patents were counted in the region of operation. For example, Alfred Teves is a German subsidiary of the U.S.-headquartered ITT. Teves' patents have been counted in the European region.

Patenting by supplier firms was estimated by developing a list of major automotive suppliers headquartered in the three principal regions, using the following sources:

Japan: Dodwell Consultants, *The Structure of the Japanese Autoparts Industry*, Tokyo: Dodwell, 1986

North America: Elm International, *The Elm Guide to Automotive Sourcing, 1987–88*, East Lansing, Michigan: Elm International, 1987

Europe: PRS, *The European Automotive Components Industry 1986*, London: PRS, 1986

This list was then compared with data on patents by company, provided by the U.S. Office of Technology Assessment. Adjustments were made to exclude nonautomotive patenting by large multi-product firms, such as Allied Signal in the United States and Hitachi in Japan.

*Source:* Estimated by the Science Policy Research Unit of the University of Sussex from data supplied by the United States Office of Technology Assessment, Washington, D.C.

## A NEED FOR EPOCHAL INNOVATIONS?

So far we've talked about innovations that involve the introduction in production vehicles of ideas already fairly well understood on the technical level. We've listed a number of advances of this type in the 1980s, and many more will be available in the 1990s—in particular, the application of electronics to mechanical vehicle systems such as vehicle suspension and the availability of mobile communications at lower cost in a much wider variety of vehicles. But what about epochal innovations—really big leaps in technological know-how such as would be entailed in workable fuel-cell power units or all-plastic body structures or sophisticated navigation and congestion-avoidance systems? As we will see, the 1990s may prove a time for such innovations. Can lean producers respond to these much more daunting challenges?

In fact, the world auto industry has lived during its first century in a benign environment—demand for its products has increased continually, even in the most developed countries; space has been available in most areas to expand road networks greatly; and the earth's atmosphere has been able to tolerate ever-growing use of motor vehicles, with minor technical fixes in the 1970s and 1980s designed to solve smog problems in congested urban areas. Shortly, the environment for operating motor vehicles may become much more demanding.

Demand for cars is now close to saturation in North America, Japan, and the western half of Europe. A small amount of incremental growth will be possible in the 1990s, but by the end of the century producers in these markets will need to provide consumers with something new if they want to increase their sales volume (measured in dollars or marks or yen rather then units). Moreover, the growth of vehicle use and increasing resistance to road building have made the road systems of these regions steadily more congested, gradually stripping motor-vehicle use of its pleasure.

New electronic-vehicle technologies that permit vehicles to navigate around congestion and even, some day, to drive themselves might solve both problems: Handing driving over to the computer would permit car companies to charge much more per car, even if they did not sell more units—and in-vehicle entertainment systems could be moneymakers as well if drivers were relieved of the need to watch the road.

Meanwhile, in the 1990s cars and trucks that are able to gather information from the roadway about congestion and to find the fastest route to their destination could make much better use of limited highway space. Given the size of the potential prize, it's no surprise that governments and motor-vehicle companies in North America, Western Europe, and Japan have recently initiated publicly funded cooperative research programs in each region to find technical solutions to these problems.[13]

However, making these technologies a reality is truly daunting. The computer industry is still a long way from adequate computing power for autopilots, and the reliability of such systems would have to be very high. Although motor vehicles under human control kill more than 100,000 people each year in North America, Western Europe, and Japan combined, it's hard to imagine that the public would accept a computer-controlled system that killed a half or even a tenth as many people.

What's more, solutions will have to be sought far beyond the research labs of individual companies, both because publicly owned roads will be a key element in the necessary information systems and because the standards selected may have a major bearing on the health of national motor industries. The recent debate over worldwide standards for high-definition television—in which governments in each major region have jockeyed to provide an advantage for their home team—are perhaps a foretaste of what is to come in the motor-vehicle industry.

A breakthrough in navigation and autopiloting could revitalize the consumer's desire to spend discretionary income on motor vehicles, even in the most saturated markets. However, even more startling breakthroughs in motor-vehicle technology may be needed simply to preserve what society has already, if the worst predictions about the greenhouse effect are borne out. These focus on the potential effects of the rising levels of carbon dioxide (partly from motor vehicles), methane, and chlorofluorocarbons (partly from auto air conditioners) in the earth's atmosphere. These emissions may dramatically raise temperatures and alter the global climate if they're allowed to continue.

In the worst case, early in the next century we may see a dramatic rise in sea levels as the Antarctic ice melts, flooding much of the world's coastal plains where population is concentrated. We may also see rainfall patterns change to convert the world's breadbaskets to dustbowls. Even much more modest changes could threaten the earth's ability to support its current population.[14]

Currently, the scientific debate on the greenhouse effect is extraordinarily confused. Everyone agrees that carbon dioxide, methane, and chlorofluorocarbon levels are rising, but the precise consequences of the increase are far from clear. Computer models incorporating the many feedback loops in the climate system are the keys to prediction. So far, however, the models are in only loose agreement and make their predictions only within a broad range. What's more, the fate of specific regions as climate changes is even less clear.

On the other hand, society is now pouring enormous scientific resources into finding precise answers, probably within the next few years. We would be surprised if the motor-vehicle industry does not have to respond in dramatic ways—and its response may provide the final test of lean approaches to research and development. For example, in the extreme case, emissions of carbon dioxide might have to be eliminated altogether, creating the need for hydrogen-powered cars, which produce only water as an end product of combustion, or even solar-powered vehicles.

So far, the Japanese lean producers have not failed at epochal innovations of this sort; they simply haven't tried, occupying themselves instead with a brilliant scavenging process that has scanned the technological landscape for ideas nearly ready for the market, as in the case of the high-tech four-cylinder engines of the 1980s. A much more difficult challenge probably lies just ahead.

# COORDINATING THE SUPPLY CHAIN

## 6

The modern car is almost unimaginably complicated. As we noted, a typical model is made up of more than 10,000 parts, each of which must be designed and made by someone. Organizing this enormous task is probably the greatest challenge in manufacturing a motor vehicle. Yet it is the one least understood and appreciated by the outside world.

Henry Ford thought he had solved the problem by the time of World War I: Do it all yourself in your own company. However, his solution raised as many questions as it answered: How do you organize and coordinate hundreds of thousands of employees in hundreds of factories and engineering offices? What do you do with your machines and factories, all dedicated to making specific parts for your own products, when demand shifts or the economy goes sour?

In the 1920s, Alfred Sloan found one answer to these problems: Do it all in your own company, but set up decentralized parts-making divisions as independent profit centers—for example, Harrison Radiator, Saginaw Steering, AC Spark Plug—to make specific categories of parts for the whole company. By treating the divisions as independent businesses, Sloan thought,

This chapter is based primarily on the research of Toshihiro Fujimoto and Richard Lamming.

## COORDINATING THE SUPPLY CHAIN

he could impose the cost and efficiency discipline of the market while preserving the coordination advantages of a unified company.

Sloan had a solution to the problem of the cyclical car market as well: When the market slumps, lay off workers in the supply system just as you lay off workers in the assembly plant.

By the 1950s, the Ford Motor Company under Henry Ford II had a new idea, which, as it turned out, was actually an old idea. Ford put out to bid to completely independent supplier firms many categories of components formerly supplied from within the company. The suppliers were given detailed drawings of the necessary parts and asked for their price per part. The lowest bidder generally won a one-year contract. When the market slumped, these suppliers were laid off by means of canceled contracts, just like workers. This was, in fact, the world Ford had left around 1913; the world of arm's-length, market-based, short-term interactions with independent businesses.

In the 1980s mass-production companies around the world were using both approaches. GM was the most integrated, with about 70 percent of the parts in each car and truck supplied by its in-house parts divisions. Saab, at the opposite extreme, made only about 25 percent of its parts, always preserving for itself just those parts most visible to the consumer—the body and the engine.[1]

How much each company actually integrated was a function of the company's history and its size. The enormous investment GM had sunk into its parts operations made it very hard to think about alternatives, while Saab was simply too small to make all its own parts. (Indeed, a key justification of the Ford purchase of Jaguar and of the GM joint venture with Saab has been that Jaguar and Saab can obtain cheaper parts through the greater bargaining power of a larger producer and can share some common parts, such as switches and lights, with Ford and GM, respectively.) However, neither system—in-house or arm's-length—works very well.

In the mid-1980s, in the twilight of mass production, many companies, including General Motors and Chrysler, experimented with reducing the fraction of parts they obtained from their in-house suppliers. This tactic was inspired by a belief that lower wages in outside supplier companies were the competitive secret of the Japanese supply systems.

In our view, this change in direction—now on hold at Chrysler and GM because of resistance by middle management and the UAW—largely misses the point.[2] The key to a competitive parts-supply system is the way the assembler (for example, Ford or Renault or Toyota) works with its suppliers (for example, Ford's Automotive Trim Operations or Bendix, Renault's Transmissions Division or Valeo, Toyota's Engine Division or Nippondenso).

Whether the supplier comes from inside the company or out makes surprisingly little difference. To see why this is so, let's pick up where we left off in Chapter 5 and follow the components supply process as it has worked (and in many cases still works) in mass-production car companies. We'll trace the system from the moment it begins in the design of a new car.

## MATURE MASS PRODUCTION: DESIGNING THE PARTS

Remember that the design process in a mass-production company proceeds in a sequence, one step at a time. First, the overall concept for the new model is specified by the assembler's product-planning team and reviewed by senior management. Then the product is planned in detail, down to the fraction of an inch (for example, the wheelbase and track) and the specific type of material to be used for each part (for example, steel fenders, plastic steering wheel, aluminum engine). Next, detailed engineering drawings are made for each part, specifying the precise materials to be used (steel of a given gauge with double galvanized coating for the fenders, for example; thermoset plastic with carbon fiber reinforcement for the steering wheel; specific aluminum alloy for the engine block, and so forth). Only at this point are the organizations that will actually make the parts called in. These typically number between 1,000 and 2,500 for the complete car, including independent companies and in-house parts divisions.

When the suppliers—whether in-house or independent—finally get the call, they are shown the drawings and asked for bids. For example, they're asked, "What will the cost per steering wheel be for 400,000 steering wheels per year?" The mass-production assembler also sets a quality target, so many bad parts per 1,000 as the acceptable upper limit, and a delivery schedule—perhaps one or two times a week—with a penalty for failure to deliver on time or in the right quantity. The term of the contract is generally quite short—usually a year for parts that require new capital

# COORDINATING THE SUPPLY CHAIN 141

investment, but even less for those commodity parts, such as batteries or tires, that most companies in the industry buy from the same suppliers and that are already in production. So price, quality, delivery reliability, and contract length become the four key elements of the assembler-supplier relationship.

When the suppliers see the drawings, they know from long experience that they are involved in a complex game, where none of the real rules are written into the bid tender. They realize that the assembler's procurement office is under tremendous pressure to reduce costs. "Cost comes first" is the assembler's byword. So quoting a low price per part is absolutely essential to a winning bid. However, the suppliers are also aware that follow-on business for a new model can often extend for ten years. Then there's the market for replacement parts, which may be considerably longer. So, in reality, they are not bidding on a one-year contract but, potentially, on a stream of business running for twenty years.

Since this is the case, should they bid below cost? Doing so is tempting, because the suppliers' experience also tells them that once a part is in production, with acceptable quality and delivery performance, they may be able to go back to the assembler for a cost adjustment: "We can't get our steel in the net shape we need, so scrap costs are running above our estimates," they may say (meaning, that the only steel blanks they can get are too big, so they must cut off more to get the size of part they need, a process that entails extra initial cost and waste), or "Our union is insisting on work-rules changes that are increasing our costs" or "The new machine we bought to mold steering wheels cannot provide adequate quality without hand finishing."

In addition, there exists a long tradition of the annual cost adjustment for follow-on contracts, designed to allow for general inflation. The assembler tends to grant these adjustments across the board without investigating individual circumstances. To look into each case would simply require too much effort. Of course, suppliers are almost certain actually to reduce production costs over time, since, as time goes on, they gain experience in producing the part. So the annual cost increase in subsequent years may turn a money-losing initial bid into moneymaking follow-on business.

Finally, for some parts requiring heavy investments in new production tools, the assembler may find it extremely costly and inconvenient to obtain a new supplier once production is in full swing. Suppliers of these parts may gamble that their ability to raise prices will grow over time. This mind-set makes the temptation to "buy the business"—that is, put in a deliberately lower bid in order to get a foot in the door—almost irresistible.

The mass-production assembler has played this game thousands of times and fully expects the successful bidders to come back later for price adjustments. Thus it is important for the assembler's product designers to have some idea of suppliers' real costs, so they can accurately estimate price adjustments downstream.

It's hard work, though. A key feature of market-based bidding is that suppliers share only a single piece of information with the assembler: the bid price per part. Otherwise, suppliers jealously guard information about their operations, even when they are divisions of the assembler company. By holding back information on how they plan to make the part and on their internal efficiency, they believe they are maximizing their ability to hide profits from the assembler.

Once the assembler designates the winning bidders, the suppliers set to work making prototype parts. This process is likely to uncover many problems, because the traditional mass-producer farms out the many parts in a complex component to many suppliers who may have no direct contact with each other. For example, until recently General Motors built practically all its own seats by ordering about twenty-five parts per seat from as many suppliers. When the parts were finally put together in the finished seat, it was not surprising that a piece wouldn't fit or that two abutting materials would prove incompatible. For example, they might rattle or squeak in cold weather because of different expansion coefficients.

Once the supplier tests the parts in components and the assembler, in turn, tests the components in complete vehicles, the assembler specifies necessary changes in each part and gives the sign-off to begin volume production. However, the mass-production assembler is still not through with the supplier-selection process.

### MATURE MASS PRODUCTION: SUPPLYING THE PARTS

At this point, the purchasing department is worrying less about getting the vehicle into production and more about how to control

# COORDINATING THE SUPPLY CHAIN 143

costs among suppliers whose operations it only vaguely understands. The obvious way to do so is to identify additional suppliers for each part and give them the final, production-ready drawings as the basis for making bids. The suppliers who already have been selected are horrified, of course—which is precisely the idea. The initial supplier also feels cheated, because the new bidder will not have to bear the cost of fine-tuning the original drawings.

Of course, the first supplier has also played this particular game countless times before and has probably left some room in its bid to make adjustments in subsequent years, as the assembler plays two or even three or four suppliers against each other. What's more, many of the assembler's threats to seek an alternative source may turn out empty, particularly when they are directed at in-house suppliers.

Let's take the example of one of GM's in-house suppliers. We'll imagine that the program manager for a new GM product is unhappy with the in-house supplier's bid—it's too high and, in the past, the supplier had quality and delivery problems. However, no sooner does the manager identify an alternative bidder outside the company than the in-house supplier goes to corporate headquarters and explains that loss of business on this part will require an increase in the cost of similar parts already being supplied for other GM products. Why? Because economies of scale will be lost and the in-house supplier will have excess capacity.

Headquarters, always very respectful of scale-economy and capacity-utilization justifications in a mass-production firm such as GM, then has a talk with the program manager. The in-house supplier makes solemn promises to try harder to reduce costs in the future while improving quality and delivery reliability—and gets the business. In this way, the internal market, which supposedly keeps the in-house supply divisions honest, is gradually diluted. This process explains how GM managed to have both the world's highest production volume and the world's highest costs in many of its components supply divisions through much of the last decade.

At the end of the selection process, the assembler usually ends up with a single supplier for the most complex and technologically advanced components, such as engine computers. For commodity parts such as tires, three or four suppliers are often put under contract. However, designating the complete roster of suppliers and beginning volume production is only the end of the first stage of assembler-supplier collaboration on a new product.

Immediately after a new model reaches the market a lengthy process of debugging begins, entailing intense interaction between the assembler and its suppliers. Despite years of testing of prototypes, the assembler often discovers from initial consumer feedback that something is not right—either a part doesn't work at all or buyers complain that it doesn't work well.

For example, a new model's brakes work properly but squeak when they are cold. The solution? A "running change," which in this case involves substituting a new brake-pad material on the production line at the earliest possible moment. In the 1980s, many Western companies introduced thousands of running changes in the first year or two of a new model's life. Each of these may have required renegotiating the supplier contracts—which meant cost increases for the assembler.

Another aspect of debugging involves manufacturability. The assembly plant may report that workers find it nearly impossible to attach a part properly because of its design, or, perhaps, there are simply too many parts in a given area of the vehicle—ten when one should do, say. The only solution may be to redesign the part or the entire component, a step assemblers almost never took before the 1980s—because it's expensive—but one they have taken more frequently in recent years as the demand for quality has grown along with awareness of the manufacturing cost penalty of an improperly designed component over the production life of a model.

Finally, the supplier may fail to meet quality targets. Remember that the mass-producer expresses these targets as an acceptable percentage of bad parts. When the assembler finds fewer than the acceptable number of bad parts, typically by inspecting parts as they are delivered to the assembly plant, these are tossed in the waste bin or sent back for a credit. Only when the number of defective parts goes above the acceptable level does the assembler do something dramatic—send the whole shipment back, say, and refuse payment.

Even in this case, it's strictly the supplier's responsibility to find the problem and correct it. Most suppliers believe strongly that "what goes on in my factory is my own business." Assembler meddling in supplier production problems is distinctly unwelcome, because it could uncover valuable data on the supplier's operations and costs—information that the assembler could use to bargain down prices for follow-on contracts.

The supplier-assembler relationship can remain conflicted

# COORDINATING THE SUPPLY CHAIN 145

even once a new model is fully debugged. If competition is unexpectedly stiff, for example, production may never reach the planned volume. So costs increase, even as pressure grows to cut prices. The assembler may be tempted to look for sources at lower cost—even outside the companies already under contract.[4]

Suppliers who have just tooled up, and who are, in fact, selling below cost, may then be dumped for a lower bidder. This step no doubt cuts costs in the short term but reconfirms all suppliers, including the new winners, in the belief that information must be guarded from the assembler and that any trust placed in a long-term relationship is trust misplaced.

As if these hurdles weren't enough, there's the problem of fluctuating volume. As we'll see in Chapter 9, the major auto markets of the mass-producers, particularly in North America, tend to be highly cyclical. There may also be rapid shifts in the mix of products that consumers demand, even when the total number of cars and trucks sold is stable. The mass-production assembler takes the position that these shifts are unpredictable and that parts orders may have to be canceled on sudden notice. Any oversupply of parts that may result is the supplier's problem. What's more, mass-production suppliers tend to have very large stocks of both finished parts and parts in process. Suppliers, therefore, build overstock contingencies into their bids, and, in the end, the consumer pays for the erratic flow of business.

As should be clear by now, the mature mass-production supply system is broadly unsatisfactory to everyone concerned. The suppliers are brought in late in the design process and can do little to improve the design, which may be hard and expensive to manufacture. They are under intense cost pressure from a buyer who does not understand their special problems. As a result, implausible bids win contracts, followed by adjustments, which may make the cost per part higher than those of realistic but losing bidders. This process makes estimating costs accurately difficult for the assembler. Moreover, the effort to play bidders off against each other makes them very reluctant to share ideas on improved production techniques while a part is in production. In other words, they have no incentive to merge their learning curves.

At best, the typical mass-production supply system can succeed in keeping the suppliers' profits very low. The assembler's purchasing department may cite this fact as the primary evidence of its success. However, parts costs—a very different matter from supplier profits—may remain very high, and quality may prove both unsatisfactory and resistant to improvement, all because no one really communicates with anyone else. Surely there is a better way.

## COMPONENTS SUPPLY IN LEAN PRODUCTION[5]

There is a better way. Again, we travel to Japan to find it. To see just how that system works, let's return to the *shusa*-led product-development process we looked at in Chapter 5. At the very outset of product development, the lean producer selects all the necessary suppliers. The leading Japanese lean producers involve fewer than 300 suppliers in each project (compared with 1,000 to 2,500 at Western mass-producers).[6] These suppliers are easy to designate, because they are typically the companies supplying the same parts for the producer's other models and are long-term members of the assembler's supplier group. (We'll explain the nature of these groups shortly.) Significantly, they are not selected on the basis of bids, but rather on the basis of past relationships and a proven record of performance.

A third to an eighth as many suppliers are involved, compared with a mass-production company, because lean producers assign a whole component—for example, seats—to what they call a first-tier supplier. This supplier is in charge of delivering complete seats to the assembly plant. In consequence, Nissan, for example, has only one seating supplier for its new Infiniti Q45 model, while GM, in many cases, is still dealing with twenty-five suppliers providing the twenty-five needed parts to the seat-building department of its assembly plants.

The first-tier supplier typically has a team of second-tier suppliers—independent companies that are manufacturing specialists. These companies may, in turn, engage helpers in a third or even fourth tier of the supply pyramid. These latter companies make individual parts according to drawings supplied by the second-tier firm. (We looked at the historical origins of this system in Chapter 3.)

The first-tier suppliers to a lean development program assign staff members—called resident design engineers—to the development team shortly after the planning process starts and two to three years prior to production. As product planning is completed,

# COORDINATING THE SUPPLY CHAIN

with continuous input from the suppliers' engineers, different areas of the car—suspension, electrical system, lighting, climate control, seating, steering, and so forth—are turned over to that area's supplier specialist to engineer in detail. The first-tier suppliers, then, have full responsibility for designing and making component systems that perform to the agreed-upon performance specification in the finished car. The supplier's development team, with its own *shusa* and with the help of resident design engineers from the assembler company and the second-tier suppliers, then conducts detailed development and engineering.

In 1988, for example, Nisshin Kogyo, a leading Japanese brake manufacturer, had a product-development team of seven engineers, two cost analysts, and a liaison person regularly positioned at Honda's research-and-development center. The team was working on a daily basis with Honda's development engineers on the design of a new Honda car.[7]

The nature of the supply arrangements means that the assembler may actually know relatively little about certain parts or systems. We interviewed the head of one Western supplier who had recently become the seating supplier to one of the Japanese transplants in North America. To get started, he had flown to Tokyo and asked the assembler for a detailed set of engineering drawings, so he could prepare a bid. However, the assembler explained that he knew very little about the technical aspects of seating and certainly had no drawings: "These are entirely the responsibility of our two traditional suppliers of seating. You will have to ask them," he said. (In the end, the Western company formed a joint venture with one of the Japanese suppliers to supply the North American transplant.)

The lean assembler doesn't delegate to the supplier the detail design of certain parts considered vital to the success of the car, due either to proprietary technology or to the consumer's perception of the product. Leading examples of parts usually reserved for the assembler's in-house supply divisions are engines, transmissions, major body panels, and, increasingly, the electronic management systems that coordinate the activities of many vehicle systems.

Even when it comes to parts where the assembler is only loosely acquainted with the technology and totally dependent on a single outside supplier, the lean producer takes care to learn an enormous amount about the supplier's production costs and quality.

But what is it about this system that allows an interchange of such sensitive information to take place? The answer is simple. The system works only because a rational framework exists for determining costs, price, and profits. This framework makes the two parties want to work together for mutual benefit, rather than look upon one another with mutual suspicion.

Almost all the relationships between supplier and assembler are conducted within the context of a so-called basic contract. The contract is, on the one hand, simply an expression of the assemblers' and suppliers' long-term commitment to work together. However, it also establishes ground rules for determining prices as well as quality assurance, ordering and delivery, proprietary rights, and materials supply.[8]

In short, the contract lays the basis for a cooperative relationship, one that is fundamentally different from the relatively adversarial relationships between supplier and assembler in the West. Similar contracts have also been commonplace between first- and second-tier suppliers in Japan since the 1960s.

## LEAN SUPPLY IN PRACTICE

Let's now take a closer look at how the supplier-assembler relationship works in practice.

At the heart of lean supply lies a different system of establishing prices and jointly analyzing costs. First, the lean assembler establishes a target price for the car or truck and then, with the suppliers, works backwards, figuring how the vehicle can be made for this price while allowing a reasonable profit for both the assembler and the suppliers. In other words, it is a "market price minus" system rather than a "supplier cost plus" system.

To achieve this target cost, both the assembler and the supplier use *value engineering* techniques to break down the costs of each stage of production, identifying each factor that could lower the cost of each part. Once value engineering is completed, the first-tier supplier designated to design and make each component then enters into mutual bargaining with the assembler, not on the price, but on how to reach the target and still allow a reasonable profit for the supplier. This process is the opposite of the mass-production approach to price determination.

Once the part is in production, in lean production, a technique called *value analysis* is used to achieve further cost reductions. Value analysis, which continues the entire time the part is being produced, is, again, a technique for analyzing the costs of each production step in detail, so that cost-critical steps can be identified and targeted for further work to reduce costs still further. These savings may be achieved by incremental improvements, or *kaizen*, the introduction of new tooling, or the redesign of the part.

Of course, all producers—mass and lean—try to analyze costs, but lean production makes it much easier to do this accurately. Where set-up times have been honed always to require only a few minutes and where production runs are frequent, short, and uninterrupted, cost estimators do not have to wait around for days or weeks to average the performance over several production runs. They can quickly collect data that is accurate and representative. Indeed collecting the data can be left to the machine operators themselves. This makes it possible to do a complete cost analysis several times a year and to monitor progress in cutting costs accurately[9]

Obviously, for the lean approach to work, the supplier must share a substantial part of its proprietary information about costs and production techniques. The assembler and the supplier go over every detail of the supplier's production process, looking for ways to cut costs and improve quality. In return, the assembler must respect the supplier's need to make a reasonable profit. Agreements between the assembler and supplier on sharing profits gives suppliers the incentive to improve the production process, because it guarantees that the supplier keeps all the profits derived from its own cost-saving innovations and *kaizen* activities.

A second feature of lean supply is continually declining prices over the life of a model. While mass-producers assume that bidders are actually selling below cost at the outset of a contract and will expect to recoup their investment by raising prices year by year, lean producers assume—or rather know—that the price for the first year's production is a reasonable estimate of the supplier's actual cost plus profit. The assemblers are also well aware of the learning curve that exists for producing practically any item. So they realize that costs should fall in subsequent years, even though raw-materials costs and wages increase somewhat. Improvements in lean-production companies should, in fact, come much faster—that is, learning curves should be much steeper—than in mass-production companies because of *kaizen*, the continuous incremental improvement in the production process.

The question is, who realizes the savings? Again, through mutual discussion and bargaining, the assembler and supplier agree on a cost-reduction curve over the four-year life of the product, with the proviso that any supplier-derived cost savings beyond those agreed upon will go to the supplier. This is the principal mechanism in the lean-supply system for encouraging suppliers to engage in rapid and continuous improvement.

Here's an example of how the process works. Say the price of a part, such as an instrument cluster, is set at 1,200 yen the first year of production. Say, too, that through the joint efforts of the assembler and supplier, a cost of 1,100 yen is achieved in the first year. In this case, the assembler pays the supplier 1,150 yen the first year. The assembler and supplier share the profit.

Now, say the supplier, through its own efforts, comes up with another innovation that further reduces the price in the first year to 1,080 yen. The supplier would keep the balance and still receive 1,150 yen. The same process would apply in the three subsequent years.[10]

By agreeing to share the profits from joint activities and letting suppliers keep the profits from additional activities they undertake, the assembler relinquishes the right to monopolize the benefits of the supplier's ideas, benefits Western suppliers would be horrified to give up. On the other hand, the Japanese assembler gains from the increased willingness of its suppliers to come up with innovations and cost-saving suggestions and to work collaboratively. The system replaces a vicious circle of mistrust with a virtuous circle of cooperation.

Once the component is designed and production begins, additional differences appear between mass and lean supply. For one, lean production has few running changes for the simple reason that the new car or truck tends to work the way it is supposed to.

Another important difference is the way components are delivered to the assembler. It is now almost universally the practice in the best lean-production companies to deliver components directly to the assembly line, often hourly, certainly several times a day, with no inspection at all of incoming parts. This procedure is in keeping with the famous just-in-time system, the invention of Taiichi Ohno.

To make just-in-time work at all—a system in which empty

parts boxes sent from the assembler back to the supplier are the signal to make more parts—yet another innovation of lean production is essential: production smoothing. Lean production, as we'll consider in more detail in Chapter 9, is characterized by extraordinary flexibility in shifting the mix of products manufactured and doing so on only a few hours' notice. At the same time, the system is extremely sensitive to fluctuations in the total volume of cars and trucks made. These types of shifts are very difficult to accommodate in a system in which employees, because of job guarantees, are a fixed cost.

So Toyota and other practitioners of lean production work very hard at *heijunka* (production smoothing), in which the total volume the assembler manufactures is kept as constant as possible. The aggressive selling system we'll look at in the next chapter largely makes the success of *heijunka* possible in the Japanese domestic market, while in export markets the Japanese lean producers have enjoyed a cost or quality advantage for thirty years. This has made it possible to cut prices during market slumps in order to keep production volume steady.

The Japanese have another motive for practicing production smoothing: They want to ensure a steady volume of business for the suppliers. That way, the suppliers can utilize employees and machinery much more effectively than in the West, where they are constantly faced with sudden changes in the volume and mix of orders at very short notice. These unannounced changes by the assembler are the cause of Western suppliers holding unnecessary stocks; they feel they must buffer themselves against sudden surges in ordering by the assemblers. After all, if you have constantly varying orders from the assembler and also need to make prompt deliveries, there's only one solution. You must make complete parts in expectation of orders and keep lots of raw materials on hand.

In Japan, assemblers give suppliers advance notice of changes in volume. If the changes are likely to persist, the assembler will work with the supplier to look for other business. The assembler will not, as in the West, suddenly pull such activities in-house so it can keep its own staff working. In Japan, there is a commitment to share the bad times as well as the good. Suppliers are, to a considerable extent, considered fixed costs, like the assembler's employees.[11]

Of course, even the best components supply system occasionally comes up with a dud, and even for the best lean producers, zero defects is a goal rather than a reality. However, a final important difference between mass and lean supply emerges once a defect is found. In old-fashioned mass-production systems, the assembler's parts checkers at the receiving dock usually spot components problems. As we saw, when the defects are few, these parts are simply discarded or returned for a credit. When they are numerous, the whole shipment may be rejected and returned. Doing so is feasible, because the assembler typically has a week of more of parts already on hand and can easily continue production while waiting for an acceptable shipment.

The lean producer has a very different attitude. With no reserve stocks, a faulty shipment could prove catastrophic. In the worst case, the entire assembly plant with its 2,500 workers might grind to a halt. Yet, this disaster almost never happens in practice, despite the fact that parts are not inspected until they are actually installed on the car or truck. Why not?

For two reasons: The parts supplier knows what faulty parts can mean and takes pains not to let it happen. As one supplier remarked to us, "We work without a safety net, so we can't afford to fall off the high wire. We don't."[12] And, in the rare event a defective part is found, the assembler's quality-control department goes rapidly through what Toyota calls the "five why's." Both the supplier and the assembler are determined to trace every defective part to its ultimate cause and to ensure that a solution is devised that prevents this error from ever happening again.

The supplier, very likely, will have a resident engineer in the assembly plant to work on problems. If the engineer can't sort out what's wrong, the assembler's engineers take a trip to the supplier's plant. This visit, however, is not an inquisition. Rather, the trip is more like a bilateral problem-solving mission.

In the world of mass production, the supplier might flatly forbid these field visits: "My plant is my business!" is a typical supplier response in the West. By contrast, lean-supply agreements in Japan always provide for plant access by assembler personnel.[13] Let's look at what might happen when they arrive.

First, they discover that the defective part has been caused by a machine that cannot hold a proper tolerance. But the machine isn't the ultimate cause. So the team asks: "Why can't this machine hold tolerance?" The supplier's personnel report that it's because the machine operators cannot be adequately trained. The team members ask, "Why?" The supplier answers that it's because these employees keep quitting to look for other work, which means the operators are always novices. "Why do workers keep quitting?" team members then ask. The answer: "Because the work is monotonous, noisy, and unchallenging." The ultimate resolution: to rethink the work process in order to reduce turnover. This, at last, is the ultimate cause—almost always an organizational problem. Once the difficulty is fixed, it's highly unlikely the problem will recur. In the process of repeatedly working through the five why's and in trying to find process improvements that reduce costs and boost profits, lean suppliers learn enormous amounts about practical paths to better manufacturing.

## MANAGING THE RELATIONSHIP

A final feature of lean supply is the supplier associations where all the first-tier suppliers to an assembler meet to share new findings on better ways to make parts. Toyota, for example, has three regional supplier associations: the Kanto Kyohokai, the Tokai Kyohokai, and the Kansai Kyohokai—with 62, 136, and 25 first-tier suppliers respectively in 1986, Nissan has two, the Shohokai and Takarakai—with 58 and 105 suppliers.[14] Most of the other Japanese assemblers have supplier associations as well. In addition, many of the larger suppliers also have associations for their second-tier suppliers, for example, the Denso Kyoryokukai at Nippondenso.

Most of the main suppliers belong to these associations, which have been extremely important for disseminating such new concepts as statistical process control (SPC) and total quality control (TQC) in the late 1950s and early 1960s, value analysis (VA) and value engineering (VE) later in the 1960s, and computer-aided design (CAD) in the 1980s.[15]

These meetings would never be possible among mass-production suppliers. They know that sharing any findings about how to make parts cheaper with less effort will only ensure that they lose the next bidding round to their rivals—or that they will win but only with such a low bid that they can't make any money. So the process of improving production techniques becomes the job of the professional engineering societies, such as the Society of Industrial Engineers in the United States. The job gets done, but very indirectly and slowly.

By contrast, suppliers to a lean producer know that as long as they make a good-faith effort to perform as they should, the assembler will ensure that they make a reasonable return on their investment. So sharing with other group members means that the performance of the whole group will improve and every member will benefit. In other words, active participation in mutual problem-solving through the supplier group is an act of simple self-interest.

Before leaving this point, we should clear up an area of misunderstanding about lean supply—the frequent assumption in the West that all parts are "sole-sourced" in a lean-supply system—that is, every part is supplied by only one supplier. This is generally true for large complex systems that require massive investments in tools—transaxles, electronic fuel-injection systems, engine computers, and so forth—but much less so for simpler parts.

Lean assemblers do worry about how hard their suppliers will try, just as Toyota and other assemblers worry about sustaining the work pace in the assembly plant. There they stick with the seemingly antiquated continuous assembly line because it is a highly effective pacing device. To make sure everyone tries hard continually, the assemblers usually divide their parts order between two or more members of their supplier group.[16] The assemblers don't take this step to drive prices down—remember that prices are not determined through bidding but by mutual investigation between the assembler and a predesignated supplier. Rather, they do it to prevent anyone letting down on quality or delivery reliability.

When a supplier falls short on quality or reliability, the assembler does not dismiss the company—the normal method in the West. Instead, the assembler shifts a fraction of the business from that supplier to its other source for that part for a given period of time as a penalty. Because costs and profit margins have been carefully calculated on an assumed standard volume, shifting part of the volume away can have a devastating effect on the profitability of the uncooperative supplier. Toyota and other companies have found that this form of punishment is highly effective in keeping everyone on their toes, while sustaining the long-term relationship essential to the system.

Lean producers do occasionally fire suppliers, but not capriciously. Suppliers are never kept in the dark about their performance. Far from it. In fact, all the Japanese manufacturers maintain relatively simple supplier grading systems. The suppliers receive scores based primarily on the number of defective parts found on the assembly line, the percentage of on-time deliveries in the proper quantity and sequence, and performance in reducing costs.

Suppliers regularly compare their scores with those of their competitors, discuss the findings, and highlight problem areas for attention, often with the help of engineers loaned from the assembler. The scoring system is not simply a statistical exercise. It also assesses the supplier's attitude and willingness to improve. Only if there is no sign of improvement will the supplier, in the end, be fired. As one assembler purchasing agent remarked in an interview: "We will stick with any supplier as long as we think they are making an earnest effort to improve. It's only when we think they have given up that we bring the relationship to an end."

These then are the elements of lean supply. Rather than price—determined by the relative bargaining power of the two sides—as the main link with outside suppliers and bureaucracy as the chief link with in-house supply divisions, the lean assembler substitutes a long-term agreement that establishes a rational framework for analyzing costs, establishing prices, and sharing profits. It is therefore in the interest of all parties constantly to improve their performance by being completely open with one another, with neither party fearing that the other will take advantage of the situation exclusively for its own ends. The relationship between suppliers and assemblers in Japan is not built primarily on trust, but on the mutual interdependence enshrined in the agreed-upon rules of the game. However, a stable set of rules doesn't mean that anyone can slack off. Quite the opposite. It keeps everyone striving constantly to improve performance.

Because lean producers are so successful in devolving much of the responsibility for engineering and making parts to suppliers, they need to do much less themselves than in mass-production companies. Of the total cost of the materials, tools, and finished parts needed to make a car, the Toyota Motor Company itself accounts for only 27 percent. The company produces 4 million vehicles per year with only 37,000 employees. General Motors, by contrast, adds 70 percent of the value in 8 million vehicles and needs 850,000 employees worldwide to do it.[17]

Part of the difference, to be sure, is that Toyota is more efficient in everything it does. A large part of the difference, however, is that Toyota and other lean producers do many fewer things. Clark and Fujimoto found, for example, that Japanese lean producers on average do detail-engineering on only 30 percent of the parts in their cars.[18] (Detail-engineering is the process of producing detailed drawings of parts that suppliers can then use to make the parts.) The suppliers engineer the rest. During the early 1980s, by contrast, the American mass-producers were performing detailed engineering on 81 percent of their parts. Yet, these mass-producers were still dealing with three to eight times as many outside suppliers as Toyota. In other words, since the Americans were doing much more of the detail-engineering and building a larger fraction of the parts they needed within the company, we would expect that the number of outside suppliers they would require would be much smaller. Instead, the situation is just the opposite. Even more striking, they needed a much larger purchasing staff. In 1987, GM had 6,000 employees in its parts purchasing operations while Toyota had only 337.[19]

### REFORMING MASS-PRODUCTION SUPPLY SYSTEMS

We've been talking so far as if the world has two types of supply systems—mass and lean, General Motors and Toyota. In fact, mass-production supply systems no longer exist in their pure form—at General Motors or anywhere else. A decade of intensified competition and the introduction of many new technologies in automobiles have led to major changes in the way Western mass-producers are treating their suppliers. We now hear a lot of talk about trust and partnership and single-sourcing. The changes aren't just rhetorical, but they don't necessarily represent a shift toward lean supply either.

To investigate the extent of these changes, IMVP researcher Richard Lamming visited the largest components suppliers and the assemblers' purchasing departments in Europe and North America.[20] To supplement his extensive interviews, we also conducted a mail survey of suppliers in North America.[21] The findings?

Intensified competitive pressures have forced Western assemblers to look for further savings in their bills for components. Some assemblers facing a crisis—Chrysler in 1981, for example— simply resorted to across-the-board reductions in the prices they paid their suppliers. Others, however, tried to drive costs down over the long run by more fully exploiting economies of scale in

# COORDINATING THE SUPPLY CHAIN

parts production. This meant rationalizing their supplier structure and reducing the number of suppliers.

This rationalization is well under way, as shown by the reduction in the number of suppliers by every mass-producer during the 1980s, from a range of 2,000 to 2,500 at the beginning of the decade to between 1,000 and 1,500 at the end.[22] The mass-producers are trying to cut the number of suppliers to each assembly plant to between 350 and 500, and have largely reached this goal, as shown in Figure 6.1.

**FIGURE 6.1**

## Cross-Regional Comparison of Suppliers

| Averages for Each Region | Japanese Japan | Japanese America | American America | All Europe |
|---|---|---|---|---|
| *Supplier Performance:* (1) | | | | |
| Die change times (minutes) | 7.9 | 21.4 | 114.3 | 123.7 |
| Lead time for new dies (weeks) | 11.1 | 19.3 | 34.5 | 40.0 |
| Job classifications | 2.9 | 3.4 | 9.5 | 5.1 |
| Machines per worker | 7.4 | 4.1 | 2.5 | 2.7 |
| Inventory levels (days) | 1.5 | 4.0 | 8.1 | 16.3 |
| No. of daily JIT deliveries | 7.9 | 1.6 | 1.6 | 0.7 |
| Parts defects (per car) (2) | .24 | na | .33 | .62 |
| *Supplier Involvement in Design:* (3) | | | | |
| Engineering carried out by suppliers (% total hours) | 51 | na | 14 | 35 |
| Supplier propriety parts (%) | 8 | na | 3 | 7 |
| Black box parts (%) | 62 | na | 16 | 39 |
| Assembler designed parts (%) | 30 | na | 81 | 54 |
| *Supplier/Assembler Relations:* (4) | | | | |
| Number of suppliers per assembly plant | 170 | 238 | 509 | 442 |
| Inventory level (days, for 8 parts) | 0.2 | 1.6 | 2.9 | 2.0 |
| Proportion of parts delivered just-in-time (%) | 45.0 | 35.4 | 14.8 | 7.9 |
| Proportion of parts single sourced (%) | 12.1 | 98.0 | 69.3 | 32.9 |

*Notes and sources:*

(1) From a matched sample of fifty-four supplier plants in Japan (eighteen), America (ten American-owned and eight Japanese-owned), and Europe (eighteen). T. Nishiguchi, *Strategic Dualism: An Alternative in Industrial Societies,* Ph.D. Thesis, Nuffield College, Oxford, 1989, Chapter 7, pages 313 to 347.

(2) Calculated from the 1988 J. D. Power Initial Quality Survey.

(3) From the survey of twenty-nine product development projects by Clark and Fujimoto. K. B. Clark, T. Fujimoto, and W. B. Chew, "Product Development in the World Auto Industry," *Brookings Papers on Economic Activity,* No. 3, 1987, page 741; T. Fujimoto, *Organizations for Effective Product Development: The Case of the Global Motor Industry,* Ph.D. Thesis, Harvard University, 1989, Table 7.1.

(4) From the IMVP *World Assembly Plant Survey,* 1990.

This development occurred at the same time that many assemblers began to outsource the production of parts that could be produced more economically by specialist suppliers than by in-house divisions. ("Outsource," in car-industry jargon, simply means buying a part from another company rather than making it yourself.) In the early 1980s, for instance, Ford in the United States shut down its in-house wiring-harness assembly operations and gave this business to twelve outside suppliers. Later in the decade it reduced the number of suppliers to four.[23]

Assemblers can reduce the number of suppliers in three ways:

First, they can tier suppliers by assigning whole components to a first-tier supplier—seats, for example—as the Japanese do. This tack can reduce the number of suppliers from twenty-five to only one, as we saw earlier. The assembler's administrative cost for coordinating supply plummets.

Second, even without tiering, assemblers can cut the number of suppliers by reducing the parts count in components. In Chapter 4, we showed how front-bumper assemblies in a GM car contained ten times the parts of a similar assembly in a Ford car. So GM might also have ten times as many suppliers to its assembly plant. Because cars and trucks are becoming more complicated, partly because of environmental demands and partly to satisfy consumers, there will always be a race between a growing number of vehicle systems and a declining number of parts per system. For the moment, however, parts counts are falling faster, with the result that assemblers are reducing the number of suppliers.

Third, assemblers can single-source parts that previously had two or three suppliers. They can convert to single-sourcing within a traditional market context by soliciting bids, then giving all the business to the low bidder. The supplier who gets the entire business should then have greater economies of scale and, hence, lower prices. Respondents to our mail survey confirmed that single-sourcing was indeed a trend. On average, between 1983 and 1988, the number of suppliers producing a specific part for each American assembler fell from 2 to 1.5 and the number of suppliers producing the same general type of part for each assembler fell from 2.3 to 1.9.[24] (As shown in Figure 6.1, this is in sharp contrast with Japan, where multiple sourcing is the rule.)

The main reason assemblers go to single-sourcing is to get longer production runs of a single component and avoid duplication of tooling. There's a down side to single-sourcing as well, however. It leaves the assembler vulnerable to supply disruptions, as happened during recent strikes affecting Ford and Renault in Europe.

Although many observers have argued that single-sourcing is another useful technique Western assemblers can learn from the Japanese, we've already seen that this argument is both wrong and beside the point. These same observers assumed that single-sourcing in Japan led to longer-term relationships with suppliers. In fact, as we saw, Japanese long-term relationships do not depend on single-sourcing but on a contract framework that encourages cooperation.

Another change in the Western supply system is the shifting attitude toward quality among assemblers. All the U.S. assemblers have instituted a quality grading system for suppliers—not just on a shipment-by-shipment basis but for all parts supplied over a considerable period. Ford started a systematic supplier grading system, called Q1, in the mid-1980s. Q1 was followed shortly by the GM Spear program and Chrysler Pentastar. These are complex statistical systems that rank suppliers by the number of defects discovered in the assembly plant, delivery performances, progress in implementing quality, improvement programs in the supplier plant, level of technology, management attitudes, and more. The aim was to bring every supplier gradually up to higher and higher levels of performance and quality. These programs had a major impact in diffusing quality-monitoring techniques, such as statistical process control (SPC), to suppliers.

With SPC, tool operators record the dimensions of each part—or a sample of parts—produced. If they notice these dimensions straying from what they should be, they either make the necessary adjustments to the machine, or, if it's a more difficult problem, such as a machine malfunction, call for help. In theory, no defective part should be produced. Part of the Q1 program involved going one step further and sharing these SPC charts with the assembler.

Our mail survey showed that 93 percent of suppliers used SPC on all their operations in 1988, up from 19 percent in 1983.[25] The Japanese began improving quality in their suppliers through the same process, though they diffused SPC to their suppliers in the late 1950s, some thirty years ago. Obviously, mass-producers still have a long way to go. In fact, once suppliers use SPC systems for a while to identify when a machine is about to produce defective parts, find out why, and then take steps to ensure the problems don't recur, SPC becomes a routine activity for production workers, a point many Japanese companies reached in the mid-1960s.

The next step in the path to lean supply, of course, would be the sharing of detailed information on the cost of each production step, using value-analysis techniques. Ironically, General Electric originally developed these techniques in 1947, and they were enthusiastically adopted by the Japanese in the early 1960s.[26] By 1988, however, only 19 percent of U.S. suppliers were providing this kind of information to their assembler customers. This fact shouldn't come as a surprise, since no fundamental change occurred in the adversarial power-based relationship between assemblers and suppliers.

We have seen some shift toward more frequent deliveries in the West. In 1983, more than 70 percent of U.S. suppliers delivered more than a week's supply of parts at once (that is, they delivered once a week or less frequently). Today, this number has fallen to 20 percent.[27] This percentage compares with 16 percent of Japanese suppliers delivering weekly as far back as 1982.[28] That year 52 percent of Japanese suppliers were delivering daily and a further 31 percent hourly. In the United States, only 10 percent of suppliers were delivering daily or hourly combined by 1988.

The American improvement in delivery schedules, though, is not a move toward lean supply. Rather, it is an attempt to cut the amount of inventory in the assembler's plant; the supplier, instead, keeps the inventories. So the change doesn't represent a philosophical shift but simply an attempt by assemblers to shift costs to their suppliers.

Moreover, it is one thing to deliver smaller lots of parts more frequently to the assembler, but quite another to produce these parts in smaller lots, as a lean supplier would do. In fact, 55 percent of U.S. suppliers produced more than a week's supply of a part at one time before changing the tools to make another part, hardly any change from 60 percent five years earlier.[29]

Many suppliers in the survey still express skepticism about the just-in-time concept. That they do is, perhaps, not surprising given the way the concept has been used by assemblers thus far. The suppliers, with some justice, see just-in-time as a way of shifting the burden of inventories onto them. Part of the problem is that, initially, just-in-time was thought of as frequent deliveries to the assembly plant. However, as we saw in Chapter 4, just-in-time comes into its own only when it is applied to production. The discipline imposed in the plant by manufacturing small lots is one of the key steps to greater efficiency and quality in lean production.

As another sign that conditions aren't changing radically, our mail survey revealed no evidence to suggest that U.S. suppliers thought U.S. assemblers any more trustworthy than they did five years ago—although we did see some moves toward longer-term contracts. The average contract length rose from 1.2 years to 2.3 years, and the proportion of suppliers with contracts of three years or more rose from 14 percent to 40 percent.[30] At the same time, suppliers reported that the assemblers had given them little assistance in reducing costs and adopting new techniques—a finding that supports our impression that those relationships are as distant as ever.

It is true that supplier engineering, combined with long-term contracts (three to five years instead of a year or less), higher quality standards, more frequent deliveries, and single-sourcing of many components characterize a new North American supply system of the early 1990s. Don't be fooled by these developments, however, into thinking that Western suppliers have been moving toward lean supply. They have not. While many of the changes resemble what Japanese lean supply looks like from the West, nearly all have been driven by cost pressures and existing mass-production logic: single-sourcing for achieving economies of scale, just-in-time for shifting the burden of inventories, and more.

Indeed, without a fundamental shift away from a power-based bargaining relationship, it is almost impossible to move toward lean supply. If the assemblers don't establish a new set of ground rules for joint cost analysis, price determination, and profit sharing, the suppliers will continue to play by the old rules.

Confronted with this power-based relationship, the suppliers' main objective is to shift any advantages to their side. Their chief way of doing so has been to introduce new technologies and bring together discrete components into systems. Without detailed value analysis, the assembler is unable to do more than guess the price of a complex component or to play off one supplier against the other.

The incorporation of many new technologies into the automobile, such as antilock brakes, electronic engine-management systems, and plastic body parts, is giving some suppliers a greater role in designing not only discrete parts but whole systems. It's also bringing many new suppliers—giants such as Motorola, Siemens, and General Electric Plastics—into the industry for the first time. The more complex the technology, the less it fits the traditional mass-production supply systems where the assembler has the upper hand. Companies supplying technologically advanced or complex components have an opportunity to add more value or, in other words, improve their bargaining power vis-à-vis the assemblers. For many suppliers, this has been the prime motivator for moving to more advanced technologies.

## SUPPLIER PERFORMANCE

While all these changes in the relationship between suppliers and assemblers have been occurring in the United States, what has been happening to supplier manufacturing performance? How big is the gap between U.S., European, and Japanese suppliers? To answer this question, Toshihiro Nishiguchi conducted a survey of fifty-four matched components plants in Japan, Europe, and North America.[31] The results, as summarized in Figure 6.1, show that manufacturing performance among Western components companies has been no better than that of the assemblers. In other words, the performance gap we found when we compared assembly plants is mirrored in the supplier industry.

In terms of component quality, the United States was within shouting distance of the Japanese—with 33 component defects per 100 cars compared with 24 for the Japanese. The Europeans, at 62 defects per 100 cars, still lagged far behind. On all other measures, however, such as the time it took to change tooling dies, the level of inventories, the number of job classifications in plants, the degree of multi-skill working, and delivery frequency, Nishiguchi found a significant gap between parts makers in the United States and Europe and those in Japan (see Figure 6.1). In most cases, the gap was larger than in assembly—a fact that indicates that the components industry is some way behind the assemblers in adopting lean manufacturing.

However, all is not lost, as there are now at least 145 Japanese components suppliers located in North America, and many more U.S. suppliers are beginning to supply the Japanese transplants in the United States. Those U.S. suppliers who have managed to secure contracts with the transplants have an excellent opportunity to learn everything from lean manufacturing and product development to lean supplier relationships. And it can be done.

When, for example, GM's Packard Electric Division began to supply the GM-Toyota joint-venture NUMMI plant in California, its initial shipments of wiring harnesses were judged competitive on price but not on quality.[32] After discussing the situation with NUMMI, Packard stationed a resident engineer at NUMMI full-time, so that quality problems could be attacked immediately. It also sought technical assistance (in the form of three industrial engineers loaned for six months) from Sumitomo Wiring Systems, one of Toyota's traditional suppliers of wiring harnesses. The Sumitomo engineers helped Packard install the full Toyota Production System in its plant at Juarez, Mexico, dedicated to supplying NUMMI. The results of Packard's continuous efforts to learn: After eighteen months it advanced from the bottom to the top in NUMMI's supplier rating.

But there is still ample potential for misunderstanding the differences in supplier philosophies, as the following example illustrates:[33]

A U.S. supplier of a complex part won business from NUMMI and steadily improved its defect rate and delivery reliability to near perfection. Then, it put in for a large price increase, a move that seemed reasonable by Western standards, since it had proved its capabilities. However, to NUMMI's Toyota-trained purchasing staff, the request seemed a blatant example of bad faith. In the Toyota system, suppliers should never commit themselves to delivering at unrealistic prices but must be prepared instead to lower their price continually over the life of the model. This kind of misunderstanding illustrates the difference in approach between mass and lean supplier relations that has yet to be bridged.

## WESTERN EUROPE AS A HALFWAY STATION

As the supplier system changes in North America, it is coming to resemble not so much the supplier system in Japan as in Western Europe. Although the mass-production assemblers in Western Europe, as we saw in Chapter 4, are now the world's most orthodox followers of Henry Ford in their own factories, the European supply system has always differed from mass-production methods and has been somewhat closer to lean supply.[34]

That's partly because European assemblers have always been smaller and more numerous. Six companies divide the mass market with shares of 10 to 15 percent of total production, while a half-dozen specialist companies split the rest of the market. These are shown in Figure 6.2 for the 1989 sales year.

These smaller assemblers never had the scale or the funds to contemplate doing everything themselves, as Henry Ford did initially and GM very nearly did for fifty years. What's more, there have always existed a number of strong European suppliers—led by the German firm Bosch, but including GKN (universal joints) and SKF (bearings)—with a clear technical lead in certain component areas. So the tradition in Europe has always been for the large suppliers to be more talented. Rather than working to drawings, many have engineered complete components for the assemblers. Clark and Fujimoto found, for example, that while the U.S. assemblers were detail-engineering 81 percent of their parts and the Japanese assemblers were detail-engineering only 30 percent, the European assemblers were detail-engineering 54 percent.[35] The size of the leading European suppliers is shown by

**FIGURE 6.2**

**Automotive Market Shares in Western Europe, 1989**

| Producer | Market Share (%) | Sales Volume (millions) |
|---|---|---|
| Volkswagen (Audi, Seat) | 15.0 | 2.021 |
| Fiat (Lancia, Alfa Romeo) | 14.8 | 1.991 |
| Peugeot (Citroën) | 12.7 | 1.704 |
| Ford | 11.6 | 1.562 |
| General Motors (Opel, Vauxhall) | 11.0 | 1.488 |
| Renault | 10.4 | 1.392 |
| Mercedes Benz | 3.2 | .434 |
| Rover | 3.1 | .412 |
| BMW | 2.8 | .377 |
| Volvo | 2.0 | .266 |
| Japanese | 10.9 | 1.457 |
| TOTAL | 100.0 | 13.478 |

Source: *Financial Times,* January 22, 1990

the fact that the European components market is the largest in the world and the top twenty companies account for one-third of the total sales of components to the assemblers. In the slightly smaller U.S. market, by contrast, the top thirty companies account for a third of total component sales.

An additional feature of the European supplier industry that makes it more similar to lean than to mass supply is the grouping of suppliers around their home-country assemblers, both physically and in terms of long-term relationships. The French assemblers, for example, have historically drawn on French suppliers, concentrated in the Paris area, with whom they have worked for decades.

What has been distinctly unlean about the European supply system is the large number of suppliers to each assembler—between 1,000 and 2,000. These large numbers indicate that lean tiering has not been the pattern, and the European mass-market assemblers are now working very hard to reduce the complexity of their supply systems by designating suppliers for whole components. This development is occurring just as Europe itself is moving from a set of national motor-vehicle production systems to a truly regional system, so the level of structural reorganization under way, even in the absence of pressure from lean producers, is substantial.

We are seeing a move toward tiering in Europe. Renault, for example, is designating the components of the car into 150 "fam-

FIGURE 6.3

**Estimated Number of Components Suppliers in North America and Western Europe**

|  | Major | Minor |
|---|---|---|
| North America | 1,000 | 4,000 |
| West Germany | 450 | 5,000 |
| France | 400 | 1,500 |
| United Kingdom | 300 | 1,500 |
| Italy | 250 | 1,000 |
| Spain | 50 | 500 |
| Other countries | 50 | 500 |
| Western Europe total | 1,500 | 10,000 |

Source: Richard Lamming, "Causes and Effects of Structural Change in the European Automotive Components Industry," IMVP Working Paper, 1989, p. 13.

ilies" of component types, Peugeot into 257, and Fiat into 250.[36] For each of these families, the companies are looking for either two or three suppliers who can deliver the complete system. Or they are attempting to bring together the suppliers of the individual parts within a system and have them cooperate in delivering a completed system to the assembly plant. This is a variant of tiering and could reduce the amount of paperwork involved in specifying the component by 50 percent. One French supplier estimates that this grouping together of suppliers can reduce as many as fifteen into one.

In the last few years, suppliers have begun to take the initiative toward restructuring the industry on a European basis. A number of suppliers have taken over companies in another country to create truly European companies supplying customers across Europe. Examples include the takeover of Jaeger and Solex in France and Lucas's electrical businesses by Magneti Marelli of Italy, and the many acquisitions by the Valeo group of France. A number of suppliers have built new plants elsewhere in Europe. Bosch and other German companies have, for instance, set up plants in the U.K. to get away from the very high costs of producing in Germany and to sell to the Japanese transplant assembly operations.

Because of the greater strength of many European components suppliers, particularly in Germany, we do not expect as many Japanese suppliers will come to Europe as have come to North America.[37] Of those who *have* come, only 40 percent have established wholly owned operations, compared with 64 percent in North America. The declared strategy of many Japanese suppliers is instead to enter into joint ventures with European suppliers. Japanese assemblers have also said that they anticipate that they'll have an easier time finding locally owned suppliers in Europe than they did in North America. That's because the Japanese think that the existing European suppliers are much better than the existing ones in the United States and Canada, so they feel they can work with the Europeans.

However strong the European components industry may be, it still faces much restructuring over the next decade. As we saw earlier, it lags as far behind the Japanese as the European assemblers do in terms of manufacturing performance and quality. Close relationships with the European assemblers may well undergo strain, as the assemblers struggle to close the gap with the incoming Japanese assemblers. Through their operations in the

United States, many European suppliers have foreseen what is to come in Europe and what must be done to close the gap. Moreover, through joint ventures with Japanese suppliers and experience in supplying parts to incoming Japanese assemblers, the European suppliers could, in fact, lead the assemblers toward lean production in Europe.

## THE REMAINING HURDLES TO LEAN SUPPLY

The Western mass-producers are now well on their way to creating a new post-mass-production supply system, which consists of the following features:

- Larger and more talented first-tier suppliers who will engineer whole components for the assemblers. They will supply these components at more frequent intervals under longer-term contracts.
- Much higher quality standards.
- Much lower costs.

Unfortunately, as we saw earlier, the reforms made to date have involved pushing the traditional mass-supply system to its limits under pressure, *rather than fundamentally changing the way the system works*. Progress toward lean supply remains blocked by the unwillingness of Western assemblers to give up the power-based bargaining they have relied on for so long. In our interviews with Western assemblers and suppliers we found strong evidence that everyone knows the words of the new song but few can hold the tune.

The fundamental problems are inherent in the system's incentive structure and logic. Many Westerners still believe that the relationship between assemblers and suppliers in Japan is based on partnership and trust alone. If only we could re-create these qualities in the West, these people say, we would make big strides toward catching up with them in efficiency. In fact, we find no evidence that Japanese suppliers love their assembler customers any more than suppliers do in the West.

Instead, they operate in a completely different framework that channels the efforts of both parties toward mutually beneficial ends with a minimum of wasted effort. By abandoning power-based bargaining and substituting an agreed-upon rational structure for jointly analyzing costs, determining prices, and sharing profits, adversarial relationships give way to cooperative ones. Cooperation does not mean a cozy relaxed atmosphere—far from it. As we saw, Japanese suppliers face constant pressure to improve their performance, both through constant comparison with other suppliers and contracts based on falling costs. However, they also have much greater discretion than in the West, with greater responsibility for designing and engineering their own products.

At the end of Chapter 4, we made the important distinction between the mind-numbing tension of mass-production work and the creative challenge of constant improvement in lean production. Very much the same contrast exists in the supplier systems. In mass production, suppliers are constantly frustrated as they try to guess the assembler's next move. In lean production, suppliers don't constantly have to look over their shoulders. Instead, they can get on with the job of improving their own operations—with the knowledge that they will be fairly rewarded for doing so.

How can the Western post-mass-production supply system move toward true lean supply? We suspect that the key means will be the creation of lean-supply systems in the West by the Japanese producers, a topic we'll return to in Chapter 9. The Japanese move will force the Western assemblers and their suppliers to go the final mile.

# DEALING WITH CUSTOMERS

**7** We've now walked through the steps in the motor-vehicle production process—the factory, research and product development, and components supply. In each of these areas, we found a large gap between the methods and results of mass production and those of lean production. The last stop in our journey takes us to the real reason for these production efforts: the consumer. We take a look at how the production system knows what the customer wants and how he or she goes about buying and maintaining an automobile. We also examine how the manufacturer goes about delivering the car to the customer.

Why didn't we begin our odyssey with the link between the customer and the production system? This might seem the logical place to start in understanding any market-driven manufacturing process. Here's the reason: Throughout this volume, we've examined each step of the production process by beginning with the perspective of mass production. And, as we've shown in earlier chapters, the success of mass production has been so geared to the needs of the manufacturing and design processes that the customer has tended to come last. So that's the sequence we've followed as well.

This chapter is based on the research of Daniel Jones, Jan Helling, and Koichi Shimokana.

## THE MASS-PRODUCER AND THE CUSTOMER

Henry Ford knew how to deal with the customer. He let the dealer do it. And, he knew how to handle the dealer as well: Keep him small, isolated, and under a binding contract to sell only Ford products. Finally, have the dealer buy cars from the factory in advance of sales, in proportion to his or her geographic sales area. That way, the dealer would have a stock of products on hand to satisfy the walk-in customer.

In practice, this last feature of the system had enormous advantages for Ford; it provided a buffer of finished units to cushion fluctuations in sales volumes. What's more, Ford demanded full payment from his dealers at the shipping dock but bought his parts and raw materials on consignment. So he could run the business with no cash invested in inventory. He was paid by his customers (the dealer, of course, not the ultimate consumer) before his bills for materials came due. If dealers balked at his efforts to force cars on them in advance, as they sometimes did in economic downturns, Ford fired his ultimate weapon: He canceled their franchise.

Perhaps this system was for the best in Ford's day. He offered only a single product, so custom orders were beside the point. The consumer might as well buy from inventory. Besides, most buyers had the mechanical skills to do their own repairs and, if necessary, could order parts directly from the factory. They had little need, as we do today, for an attentive dealer (or mechanic) to keep their car running.

But Ford's system set a bad precedent. It clearly signaled that the production needs of the factory came first; the dealer and the customer were expected to make any necessary accommodations. As product offerings became more varied under Alfred Sloan and cars became steadily more complex mechanically, Ford's approach to customer relations became less and less satisfactory.

The system changed very slowly, however. In the late 1940s, the U.S. Supreme Court took away the right of assemblers to enforce exclusive selling clauses in their dealer franchises. These clauses gave the factory the right to cancel the franchise of any dealer who attempted to sell a competitor's products through the same dealership.

The demise of exclusive dealing meant nothing at the time.

# DEALING WITH CUSTOMERS

The industry was already dominated by the Big Three auto makers, which had established their own dealer chains, a situation that wouldn't change for at least another decade. Theoretically, the Supreme Court decision might have helped tiny companies, such as Nash and Studebaker, to survive. But these were on their way out anyway, so the decision had no real impact on them either. However, once imports began to arrive on the scene in the late 1950s, many of the smaller and weaker dealers began to "dual"—that is, to add a second or third line of cars from a foreign assembler in an attempt to beef up their business.

For new competitors in the U.S. market, such as Volkswagen and Renault, the ability to pick up existing dealers quickly and inexpensively paved the way for rapid growth in market share. For example, in only two years—from 1957 to 1959—the import share of the U.S. car market jumped from 2 to 10 percent, a rate of increase that would have been (and still is) quite impossible in Europe and Japan, as we'll see shortly.

There have been other changes in U.S. car dealerships over the decades. As the investment in equipment necessary to service cars increased, the number of car dealers gradually dropped. From 45,500 in 1947, the number fell to 30,800 in 1970, and to 25,100 in 1989.[1] It will probably drop to less than 20,000 during the 1990s.[2] As the car market grew and the numbers of dealers fell, the number of cars sold per dealer increased from 70 in 1947 to 393 in 1989 (580 if light trucks are included), as shown in Figure 7.1.[3] Recently, the Japanese and Korean producers have been cautious about expanding their dealer networks and have instead pushed up the number of cars sold per dealer to record levels, as shown in Figure 7.1. Also, in the past few years a number of "megadealers" have emerged to challenge traditional single-site dealerships. These entrepreneurs maintain forty or more outlets and, in many cases, sell a dozen or more brands.

In other ways, however, the dealership system has hardly changed at all since Henry Ford's day. Overwhelmingly, dealers are still small, individually owned businesses. Some 11,700 of them, or 47 percent, are still single-site operations. In many cases, they still pay cash up front for cars and still complain about the assemblers forcing them to take cars they don't want. Inventories are still large—averaging sixty-six days' stock on hand over the last decade,[4] just above the sixty days that industry considers the optimum level, where plenty of cars are on hand to provide variety for the walk-in buyer but carrying costs are not excessive.

### FIGURE 7.1

**Car Sales per Dealer in the United States**

| Producer | 1956 | 1965 | 1978 | 1987 |
|---|---|---|---|---|
| General Motors | 183 | 351 | 464 | 249 |
| Ford | 189 | 318 | 389 | 259 |
| Chrysler | 104 | 213 | 239 | 114 |
| Honda | | | 396 | 693 |
| Toyota | | | 423 | 578 |
| Nissan | | | 323 | 477 |
| Hyundai | | | | 1369 |
| Volkswagen | | | 253 | 219 |
| Volvo | | | 120 | 257 |

*Source:* Compiled from *Automotive News Market Data Book*, various years

In some ways, the dealership system has even moved backward since Ford's day. Every assembler maintains an enormous marketing division for each of its sales divisions (for example, Chevrolet, Mercury, Dodge) with a main office near headquarters and regional offices to oversee dealers in each geographic area. The marketing division and the dealers typically have strained relationships, because the division sees its job as making sure that the dealers sell enough of the cars the factory must make to maintain steady production. The sales division's key activity is to juggle incentives for both consumers and dealers so all the cars are sold.

To achieve this goal, the marketing division may tie dealers' orders for popular cars to their acceptance of unpopular cars—a highly effective but extremely unpopular method of meshing supply with demand. For example, we recently visited a divisional headquarters at a U.S. company. Headquarters was facing the problem of how to sell 10,000 already built cars that no dealer wanted. The company had built the automobiles based on its forecasts of market demand rather than on actual orders from dealers or consumers. The market had changed, however, and no one wanted the cars.

One possible solution was the one we just discussed: to tie orders from the dealers for more popular models to the slow sellers. So to get five popular models, each dealer had to accept one unwanted model. Another was to offer a factory rebate on the

# DEALING WITH CUSTOMERS

unwanted models. This option was much more acceptable to the dealers but considerably more expensive for the company.

To make matters worse, coordination between the sales division and product planners in the big mass-production companies is poor. While the product planners conduct endless focus groups and clinics at the beginning of the product-development process to gauge consumer reaction to their proposed new models, they haven't found a way to incorporate continuous feedback from the sales division and the dealers. In fact, the dealers have almost no link with the sales and marketing divisions, which are responsible for moving the metal. The dealer's skills lie in persuasion and negotiation, not in feeding back information to the product planners.

It's sobering to remember that no one employed by a car company has to buy a car from a dealer (they buy in-house through the company instead, or even receive a free car as part of their compensation package). Thus, they have no direct link to either the buying experience or the customer. Moreover, a dealer has little incentive to share any information on customers with the manufacturer. The dealer's attitude is, what happens in my showroom is my business. (In this respect, the relationship between dealer and manufacturer resembles the one between components supplier and assembler.)

As it happened, we visited one divisional manager in Detroit on the day he was first shown the production-ready version of a major new model. The manager told us that the car he saw was entirely different in character and consumer appeal from the prototype he had agreed to sell two years earlier. Since then, the sales division had had little contact with the product-development team, which made many changes to the vehicle in order to make it easier to manufacture. As far as the sales manager was concerned, however, these changes jeopardized the sales appeal of the car in the market, and now it was too late to make any adjustments. In the end, the divisional manager's judgment proved right—the product has been a disaster.

In fact, the assemblers' sales divisions have grown into enormous bureaucracies that cannot effectively communicate market demand back to product planners. Moreover, they antagonize the dealers, with whom they should have a collaborative relationship.

What's more, the bazaar tradition of car selling—where the customer and the dealer try to outwit each other on price—is still firmly in place at dealerships, even though more and more buyers report in surveys that they thoroughly dislike it. And that means the flow of information between customer and dealer is restricted as well.

Salespeople, that is, aren't really interested in the customer's needs or desires. They want to close the deal as soon as possible and will present only selected bits of information about the product to achieve that end. Once the deal is signed, the salesperson has no further interest in the customer. The entire selling and negotiating system is based on giving the customer as little real information as possible—the same principle on which the relationship between dealers and manufacturers is based.

The result? As the factory and engineering shop have become more efficient under the pressure of lean competitors, the after-the-factory component of the car business—a component that includes not just dealing costs (manufacturer's advertising and promotions, shipping, staff and overheads, and more) but also the dealer's advertising and warranty work—has accounted for a larger and larger fraction of the total cost the consumer pays. Most analysts currently estimate that 15 percent of the buyer's total cost is incurred after the factory gate, when the new car is turned over to the assembler's selling division before being sent on to the dealer.

As post-factory costs have risen to a larger percent of total costs, the assemblers have understandably begun to focus more attention on pushing these costs down. However, studies of retailing more generally show that auto distribution costs in North America and Europe are already considerably lower, as a fraction of total cost, than for many products, including food. In fact, what the distribution system already offers is low cost, but with an even lower level of service.[5]

We can see other elements of Henry Ford's dealership system still firmly in place today. In the 1980s, the elimination of special ordering became a favorite method for mass-producers to try to improve efficiency in their factories and supply chains. With special ordering, a customer would go to a dealer and specify a car with a particular set of options. The car would be built once the order was placed. Custom ordering a new car was once an annual or biannual ritual for many Americans and Canadians but is now much less common. As the retiring general manager of a sales division at a Big Three company recently told us with some satisfaction, "If I've accomplished nothing else in my years here, I have succeeded in stamping out special orders!"

The European and Japanese exporters to the United States have never accepted special orders, because of the distances involved in supply. Rather, they concentrated on adding a variety of options as standard equipment on the cars they export. With the growth of import franchises over the years, consumers have many more choices. In 1958, for example, U.S. consumers could buy twenty-one different makes of car from ten different manufacturers; in 1989 they could buy 167 different models sold under thirty-seven different makes from twenty-five different producers. So the North American consumer now has an enormous and growing variety of products to buy off the dealer's lot. However, if a vehicle isn't already available that fits the customer's requirements, he or she may find it difficult to special order one.

## The Customer in Europe

The European distribution system has, in many ways, closely resembled the American, except in many respects it's thirty years behind. In Western Europe, there are not only more dealers than in the United States, but in many countries there is still a two-tier dealer structure, something that disappeared from the United States in the 1930s. That is, in addition to Europe's 36,200 main dealers, there are also 42,500 subdealers.[6] Most of these are small repair shops selling new cars that are supplied through the main dealer acting as a wholesaler. Compared with the United States, where each dealer sells, on average, 393 cars a year, the average main dealer in Western Europe sells only about 280 cars a year, and, if we count the subdealers, only 128 cars a year each. (Data comparing sales per dealership in the United States with Europe and Japan are given in Figure 7.2.)

Europe's system also has the additional complication of another layer between the manufacturer and the dealer, namely, the national import company, which performs many of the functions—such as overseeing dealers—of the regional sales office in the United States. However, in many cases these companies are not owned by the manufacturer. For example, Volvo cars are sold in the U.K. through Volvo Concessionaires, which is owned by the Lex Group—a company that also owns many car dealers in the United Kingdom.

### FIGURE 7.2

**Car Sales per Dealer, by Region, 1984**

|  | Main Dealer | Main & Sub-Dealers | Dealers of Domestic Brands | Dealers of Imports |
|---|---|---|---|---|
| United States | 355 | na | 396 | 225 |
| Europe: |  |  |  |  |
|   United Kingdom | 321 | 233 | 359 | 148 |
|   West Germany | 189 | 119 | 192 | 59 |
|   Italy | 339 | 111 | 220 | 64 |
|   France | 325 | 61 | 58 | 80 |
| Japan | 222 | na | 222 | na |

*Note:* The United States and Japan do not have sub-dealers. There were virtually no imports into Japan in 1984. "Main & Sub-Dealers" is the average of both.

*Source:* SRI International, "The Future of Car Dealerships in Europe," 1986; *Automotive New Market Data Book;* and K. Shimokawa, "The Study of Automotive Sales, Distribution and Service Systems and Its Further Revolution," IMVP Working Paper, May 1987, p. 9.

Moreover, in the last decade, the number of main and sub-dealers has actually increased in most of Western Europe. And, with the opening up of the French, Spanish, Italian, and Portuguese markets to the Japanese during the 1990s, the number of dealers could increase even further—even though some of the new dealers in Japanese cars in those countries will have previously sold other existing makes.

The only exception to this trend in Europe is the United Kingdom, where the number of dealers has steadily dropped—from 12,000 in 1968 to 8,144 in 1988.[7] In some respects the car-retailing structure in the United Kingdom has gone even further than in the United States toward the megadealer, with the growth of large publicly owned dealer groups owning many dealerships selling a variety of makes on separate sites. The largest of these publicly owned dealer groups, such as Lex, are now expanding into the United States and into continental Europe.

It is not only the structure of the distribution system in Europe that differs from that of the United States. The legal framework does as well.

The European assemblers have never legally lost their right to enforce exclusive dealership clauses in their franchises, so market access for new imports has always been much more difficult than in the United States. If you've ever wondered why Japanese cars were for years sold out of garages on back streets in most European countries, you don't have to look much farther for the answer than the fine print in the contracts of European franchises. In most cases, the franchises prohibit the selling of another brand on the same site—making it difficult for the Japanese to find dealers. Moreover, the Japanese have been constrained by quotas in many European markets, so the volume they can offer is not always enough to tempt a larger, established dealer to switch franchises to them, although this situation is now changing.

So far, again with the exception of the U.K., there has not been a shakeout of the weaker dealers in Western Europe. Most of the subdealers still sell cars made by the local producers. As competitive pressure leads to a rationalization of the dealer network, many of these dealers are the first to go. This happened in the U.K. in the 1970s, as Rover cut its dealers from 6,800 in 1968 down to 1,900 in 1982 in response to its falling share of the market.[8] This move proved a golden opportunity for the importers who were just entering the U.K. market to snap up these dealers and create their own network. We can expect to see a similar rationalization of the dealer networks in France, Germany, and Italy, as the European car market becomes more integrated, and competitive pressures from the Japanese mount. So far, however, the Japanese have faced the very expensive task of starting distribution channels from scratch.

A major question for the newly unified Europe will come in 1995, when the current exemption of car dealers from restrictive selling laws comes up for review. In the case of almost every other consumer product, the European Community requires a factory to permit its franchisees to sell competing brands. In 1995, the Community must decide whether to follow U.S. or Japanese or traditional European practice in organizing its distribution network for cars. As we've seen, this is as much a trade issue as one of competition policy, because the ability of dealers to "dual" will almost certainly ease the importers' path.

In addition to a more complex structure, the European system exhibits the same inefficiencies as the American. European dealers' finished-unit inventories are similar to those in North America. Moreover, the fraction of total costs that the consumer incurs after the factory gate is about the same in Europe as in North America.

One feature of the distribution system has developed differently in Europe. This is the sale of executive, luxury, and high-performance sports cars. As part of a deliberate strategy of seeking to differentiate these cars from those offered by the volume producers, some specialist producers instituted a much higher level of service to the customer. For example, Volvo, together with its U.K. importer, Volvo Concessionaires (owned by Lex Services), pioneered a lifetime service contract and other forms of enhanced care.

This approach was rapidly copied by other European specialists and has been applied in their dealerships in North America. For example, part of the successful sales recovery of Jaguar in North America in the early 1980s was through enhanced customer care, which overcame consumer concerns about the reliability of the product. Recently the new Japanese luxury brands—Acura, Lexus, and Infiniti—have pushed farther down this path by requiring that their dealers spend large sums on dedicated selling sites built to standard designs and on staff training.

European specialist producers have also continued and even encouraged custom ordering of cars. In the German domestic market, for example, Mercedes offers no option packages; rather, all options are "free standing" and installed at the factory to customer order. (The inability of the factory to do this easily or accurately is, in fact, one of the causes of low productivity in the European specialist plants we visited in Chapter 4.)

However, in other respects the basic dealer structure has remained the same, even in these luxury dealerships. It is taken as a given that a high level of service entails high distribution costs and that this can only be justified on luxury makes with high gross margins. Cheaper cars can logically only be sold through dealers offering the minimum of assistance to the consumer.

## THE LEAN PRODUCER AND THE CONSUMER

Is there an alternative, lean approach to selling and servicing cars, an approach that completes the lean-production system?

We think there is, at least in logic, and a number of its elements can be seen in Japan today. The Japanese system is not an ideal model of a lean distribution system for a number of reasons and, in fact, as we'll see shortly, it is changing. However, the way the Japanese producers think about distribution in their domestic market and the manner in which the pieces of their system fit together point toward the lean distribution system of the future, a system hardly imagined in the West.

To see lean distribution for what it should be, we can't begin from a narrow cost-cutting perspective, the normal Western approach, in which factors such as the number of sales made by each salesperson per month are the way to measure success. Rather, we must view it as an essential component of the entire lean production system.

Let's begin by visiting a typical car dealer in almost any Western country. The premises consist, effectively, of a large parking lot on which sits a vast array of new cars gathering grime and running up interest costs. The sales personnel, who work on individual commissions, live on a fixed proportion of every sale plus a small base wage. Most are professional sellers, not product specialists. That is, they've received their training in sales techniques, particularly in how to drive an effective bargain, rather than in the special features of what they're selling. So it really doesn't matter if they're selling shoes, computers, encyclopedias, or cars.

We've visited showrooms for years as part of our IMVP work and are continually amazed at just how little salespeople *do* know about their products: The salesman who defended the merits of rear-wheel drive in the front-wheel-drive car he showed us; the saleswoman who argued for the fuel economy of the four-cylinder engine in the V6 model on display; the salesman who volunteered that shoes, his previous product specialty, were a lot easier to sell than the cars he had been selling for the past two weeks—these are only a few examples of salespeople's glaring lack of product knowledge, a problem that is particularly acute in North America. (The turnover of sales personnel in Europe is much lower, and they seem to know more about the specific products they are selling.)

While it may still be possible to special order some makes of car in North America, the sales staff pushes the customer very hard to take a car already on the lot, perhaps by offering a better discount. Once a deal is struck, after some intense haggling, the customer, now the buyer, is turned over to the financial staff to arrange payments, then to the service staff to arrange delivery. The service department is in charge of taking care of any subsequent problems.

Three months after the sale, the buyer usually receives a

questionnaire from the assembler. "Were you satisfied with the car and with the dealer?" the company wants to know. And, for some years after the sale, the buyer is likely to receive a monthly or quarterly magazine from the assembler with a few general-interest articles and information on new products.

That's the extent of the relationship between buyer and seller for the most expensive consumer purchase most of us make in our lives. (Remember that our homes, the other big-ticket item in our personal consumption, usually appreciate in value, while our cars depreciate over a decade or less to near worthlessness. So, in terms of net consumption, cars are much more important than housing.)

How does the scenario we just sketched contrast with the sales practices of the lean Japanese producers? Again, let's take Toyota as an example.[9] Toyota has five distribution "channels" in Japan—Toyota, Toyopet, Auto, Vista, and Corolla, and is about to open a sixth. (Nissan and Mazda also have five separate channels each and Honda and Mitsubishi have three each.) The channel is simply the name on the dealership. In the United States, for example, we might see the name of the individual owner—Joe Smith Buick, say. In Japan, it's Toyota Vista or Toyota Corolla. The channels are nationwide and in many cases are owned by the assembler. Each channel sells a portion of the total Toyota product range. For example, one channel may sell less expensive models, another sportier ones, and so forth.

The channels have different labels and model names for their cars, but the main thing that differentiates them is their appeal to different groups of customers. Since every car in all five channels is clearly identified as a Toyota, the purpose of the channels is not to establish brand identity, as with sales divisions in the United States. Rather, it's to develop a direct link between the manufacturing system and the customer—whom Toyota, not accidentally, calls the owner.

To see just how the Toyota system works let's look at one of the five channels, Corolla. Toyota established this channel in 1961 to sell the Publica model but changed the name to Corolla in 1966 when this new model replaced the Publica in the Toyota product range. It has since expanded its lineup to include the Supra, Camry, Celica, and Corolla II models and the Townall van and pickup.

The Corolla channel is directly tied into the product-development process. During the entire period that new cars destined for sale through the channel are being developed, staff members from the channel are on loan to development teams. These channel representatives are in a position to make an invaluable contribution to product development, for reasons we'll examine in a moment.

The channel, which is part of the Toyota company, sells its cars through seventy-eight dealer firms, each with about seventeen different sites. (This contrasts with the several hundred or thousand dealer firms the mass-producer has to work with.) About 20 percent of the dealers are owned by the Corolla channel.[10] The rest are either partly owned by Corolla or independently owned, although all the training is done centrally by Corolla. Each of these dealerships has had a long and close relationship with Toyota, and they can best be characterized as part of an extended family that is fully integrated with the parent. In addition to training, the channel provides staff and a full range of services for those dealerships where it does not own the facilities. All told, the channel sold about 635,000 cars and trucks in 1989 and has 30,400 employees.

The employees, many of whom are college graduates, are hired right after graduation each spring. They undergo an intensive training program at the Corolla "University," which offers sixty courses, mostly related to marketing. Once the new employees are fully trained—although formal training continues every year for every employee—they're assigned to specific dealerships and begin selling cars.

The sales staff in each dealership is organized into teams of seven or eight, an organization very similar, in fact, to the work teams in the Toyota and NUMMI assembly plants we described in Chapter 4. Just like those in the factory, these teams are multi-skilled; all members are trained in all aspects of sales—product information, order taking, financing, insurance, and data collection (which we'll explain shortly). They're also trained to systematically solve owners' problems as they arise.

Each work team begins and ends the day with a team meeting. During the bulk of the day, team members disperse to sell cars door-to-door, with the exception of one team that staffs the information desk in the dealership. Each month, the entire team takes a day to solve systematically any problems that have cropped up, using the "five why's" and other problem-solving techniques. These meetings are the sales equivalent of the quality circle in the factory.

Selling of cars door-to-door is unique to Japan and is uniformly bewildering to foreign observers. Here's how it works. Team members draw up a profile on every household within the geographic area around the dealership, then periodically visit each one, after first calling to make an appointment. During their visits the sales representative updates the household profile: How many cars of what age does each family have? What is the make and specifications? How much parking space is available? How many children in the household and what use does the family make of its cars? When does the family think it will need to replace its cars? The last response is particularly important to the product-planning process; team members systematically feed this information back to the development teams.

On the basis of the information they've garnered and a knowledge of the Corolla product range, the sales representative suggests the most appropriate specification for a new vehicle to meet this particular customer's needs. The family may have doubts about what to buy, of course, even if it's really in the market for a car, so the sales representative may bring a demonstration vehicle on the next visit. Once a household is ready to buy, it places a special order through the sales representative. A vast majority of cars in Japan are customer ordered, just as this possibility is being eliminated in the United States. The vehicle order also typically includes a complete financing package, trade-in on the old car, and insurance, because the sales agent is trained to provide one-stop service for the auto buyer.

If cars are customer ordered, you might ask, how does the factory cope? Here's what happens.

Factory executives try to make an educated guess about the demand for different versions, colors, and so forth. On the basis of this forecast, they establish the plant's build schedule, which they also give to the components suppliers so the latter know what to make. The accuracy of these forecasts obviously depends on how frequently the build schedule is revised. This is typically every ten days in Japan, compared with every month to six weeks in the West.[11] Once the orders come in, the assembler adjusts the build schedule to make the specific cars the customer wants. Because the Japanese practice just-in-time production, doing so is much easier than in the West, which has much less flexible factories and much longer lead times for ordering parts (which sit around as inventory for a long time before they are used).

Of course, the Japanese build schedule is more accurate in

## DEALING WITH CUSTOMERS

the first place and can accommodate a customer-specified order more easily because of the much quicker feedback from the customers about what they really want, and because Japanese dealers keep a much closer watch on trends in tastes. The assembly plant and the components suppliers can plan ahead more accurately and get the right mix of products going down the line—for example, mixing some high-specification cars that take a little longer to build with low-specification cars that take less time. Japanese factories can deliver a customer-ordered car in under two weeks in Japan. The same order—if the customer could get it at all—would take six weeks in the West, at best, and could take as long as three months.

What about determining a price? Because the customer is buying a car tailored to his or her needs, the haggling that Western car buyers find so distasteful is almost eliminated in the Japanese system. The salesperson doesn't need to discount the product in order to get rid of a car that the customer would rather not have. Moreover, the prime objective of Japanese dealers is to keep the customer feeling that he or she is part of the dealer's "family." Dealers want customers to think that they have been treated well and have paid a fair price.

Remember that this is likely to be one of many transactions that the customer does with this salesperson. The salesperson will probably already have sold the customer a car in the past, taken care of the formalities of registering it and disposing of the traded car, arranged to have the car serviced, and seen it through the rigorous government inspections. Quite possibly the salesperson will also have battled with the insurance company for an accident claim on the customer's behalf and lent him or her a car while the customer's own car was being repaired. In the West the pressure is on to make the most out of a one-off transaction between two strangers with no subsequent loyalty or commitment. (Even if the customer returns to the same dealer for a subsequent purchase, the salesperson is likely to have moved on.) In the Japanese system the aim is to maximize the stream of income from a customer over the long term.

With very few defects in Japanese cars and intense competition in the Japanese car market, it is clearly understood that the dealer will fix any problems the owner encounters with the car even after the end of the formal warranty. The customer need not argue with dealers to get them to accept responsibility for warranty claims, an unpleasant experience that usually convinces

Western customers to look elsewhere for their next car, particularly to a brand that has a reputation for few flaws. Once a contract is signed, the order goes directly to the factory. When the car is ready, in ten days to two weeks, the sales representative personally delivers it to the new owner's house. The new car buyer need never go near a dealership.

### The Lean Dealership

Some Japanese buyers, particularly younger ones in big cities, would rather visit a dealer. Unlike older people, they're more interested in shopping around and seeing for themselves what products are being offered. This trend is occurring just as manufacturers are finding it increasingly difficult to recruit people willing to sell cars door-to-door. For one, the assemblers are hiring more women as salespersons, some of whom are not eager to knock on doors, particularly during the evening. The result: More and more Japanese are buying their cars at the dealership—about 20 percent do so in the Corolla channel, and this percentage is higher for other channels. In addition, as we'll see, all buyers will visit the dealer eventually to get their car serviced.

A typical modern Corolla dealership may look similar to Western dealerships in one respect—its showroom—but everything else is different. For starters, no vast parking areas exist; in fact, you'll see few cars on the premises other than three or four demonstrator models. Since most cars are manufactured to order, there is no expanse of finished vehicles to buy off the lot and no sixty- or seventy-day stock of cars running up interest costs. In Japan, the stock of finished cars in the system averages only twenty-one days.[12]

Second, a salesperson doesn't descend on the hapless shopper. Since the team is paid on a group commission, the seven or eight team members in the showroom have no incentive to grab the customer before the next salesperson does or suggest that he or she can provide a better deal. Instead, members of the team all join in the discussion once the customer approaches them to ask a specific question.

The heart of any Japanese dealership is its service area. The area's primary purpose isn't to rectify problems or carry out routine servicing as is the case in Western dealerships. Rather, its

# DEALING WITH CUSTOMERS

raison d'être is to prepare vehicles for the Ministry of Transport inspections, a task that provides a major source of revenue. (Government inspections also exist in a milder form in Europe and are truly mild, by comparison with Japan, in North America.) All cars must pass the first inspection when they are three years old. The Ministry then requires inspections every two years, until a car's tenth year, whereupon inspections are required annually.

The cost of these inspections becomes quite high as cars age. Tests not only become more frequent, but more demanding. For example, by about the seventh year the entire brake system will probably need replacement, even if it functions normally. So the Japanese have a strong incentive to buy a new car after four years, and most Japanese retire their cars at this time. (The dealer resells only a third of the trade-ins in the local market. Another third is shipped to other Southeast Asian countries for sale there, and one-third is scrapped, because the repair costs the dealer would incur to fix them to meet test standards are prohibitively high. As we'll see, apart from wear and tear, dealers fix anything that goes wrong, so they don't want to take the risk of taking on high future costs or tarnishing their reputation.) Car buyers have little opportunity to postpone their next purchase in hard times—a common response to economic downturns in the West.

## Channel Loyalty in Lean Production

We're used to the idea that buyer loyalty to a given brand is largely a vestige of the past in the Western auto markets. The fact that a customer buys a Chevrolet or a Renault once doesn't increase the likelihood of that customer's buying a Chevrolet or Renault the next time around. Far from it. Most Western consumers shop around today, looking for a good bargain or an available vehicle that meets their needs. They pay relatively little attention to specific brands.

In the U.K., for example, brand loyalty has fallen from around 80 percent in the 1960s to 50 percent today. It's even lower in the United States, where, moreover, repurchase of the same brand falls with the consumer's age—from about 30 percent for those above age fifty-six, 22 percent to 23 percent from ages twenty-six through fifty-five, and 13 percent for those under age twenty-five.[13]

This situation doesn't exist in Japan. A key objective of every distribution channel is to build and nurture lifetime channel loyalty. Again, let's look at the Corolla channel.

Once a new car is delivered, the owner becomes part of the Corolla family. This means frequent calls from the person selling the car—who henceforth becomes the owner's personal sales agent. The representative will make sure the car is working properly and ferret out any problems the owner may be having to relay back to the factory.

The sales agent also sends the owner a birthday card or a condolence card in case of death in the family and will call to ask if sons and daughters will need a car as they leave for college or their first jobs. It's often said in Japan that the only way to escape the sales agent from whom you once bought a car is to leave the country.

One aspect of this relationship would no doubt be particularly welcome to Western car buyers. Because the channel is obsessed with market share and tries never to lose a single owner, the relatively short-term warranties Japanese assemblers offer are ignored. The channel will, in general, continue to fix defective cars at no cost to the owner throughout the vehicle's normal life, provided the owners have not abused them. (Obviously, this implicit warranty doesn't apply to routine wear, such as replacing brake and clutch linings.)

## LEAN VERSUS MASS DISTRIBUTION: A SUMMARY

As we've seen, the lean approach to dealing with customers is significantly different in concept from the mass-producers' approach. First, the Japanese selling system is active, not passive; indeed the Japanese call it "aggressive selling." Rather than waiting at the dealership for customers attracted by advertising and publicly announced price cuts, such as factory rebates, the dealer's personnel periodically visit all the households in the dealer's service area. When sales lag, the sales force puts in more hours, and when sales lag to the point that the factory no longer has enough orders to sustain full output, production personnel can be transferred into the sales system. (This type of transfer occurred during Mazda's crisis in 1974 and, more recently, at Subaru.)

Second, the lean producer treats the buyer—or owner—as an

integral part of the production process. The elaborate data collection on owner preferences for new vehicles is fed systematically to development teams for new products, and the company goes to extraordinary lengths never to lose an owner once he or she is in the fold.

Third, the system is lean. The whole distribution system contains three weeks' supply of finished units, most of which are already sold.

The system that delivers this high level of service is also very different from a mass-production dealer system. The industry is very much more concentrated—there are only a total of 1,621 dealer firms in Japan, compared with some 16,300 dealer principals in the United States, a market two and a half times larger than Japan. Almost all Japanese dealers have multiple outlets and some of the largest easily match the megadealers found in the United States. In the same way as lean producers only have a limited number of suppliers, they only work with a limited number of dealers, who all form an integrated part of their lean-production system.[14]

## THE FUTURE OF LEAN CUSTOMER RELATIONS

If many elements of the Japanese system are superior, as we think they are, why haven't they been copied in the West? When we asked the Japanese lean producers and the Western mass-producers this question, two completely different answers emerged. The Western producers unanimously argue that the system is too expensive, "a cost-control nightmare they wish they could get out of in Japan," to quote one. It requires large amounts of effort to sell each car, they maintain, as shown by the fact that the average sales representative in a U.S. dealership sells ten cars per month (or about one car every two days), while the average Japanese sales representative sells four cars per month (or about one a week). From the mass-producer's perspective, in which the costs of selling are already too high, this extra cost seems completely impossible to justify.

The Japanese perspective is quite different. First the system of door-to-door selling is viewed as an anachronism suited to special conditions in Japan. It is gradually being phased out there, so it is only the other elements of the system that the Japanese would wish to introduce in North America and Europe. However, and most important, as one Japanese executive pointed out: "The system makes no sense unless cars are built to order and delivered almost immediately. We can do this only as we develop a complete top-to-bottom manufacturing system in North America and Europe by the end of the 1990s."

In the current situation, where Japanese cars are manufactured 7,000 miles and many weeks from the U.S. customer and where Japanese companies are constrained by market-share limitations in many markets, they've chosen to behave like Western mass-producers. We believe that following the Western system is not their ultimate intent, and that the Western mass-producers may be in for a final surprise before the end of the 1990s, as this last element of the lean-production system emerges.

The Japanese companies are quite aware of the costs of their system—no one is better at analyzing the costs of every step in production down to the last yen. They argue that these costs would make no sense if lean selling were serving the same functions as mass-production selling. However, they point out, it does much more. The lean selling system, with its periodic surveys of practically all consumers in the Japanese market, is the first step in the product-development system. It avoids the need for the time-consuming, expensive, and frequently inaccurate market assessment surveys of the Western mass-producers.

The lean selling system also reduces inventory costs dramatically and smooths the flow of production in the factory. By making sure its sales force has a clear understanding of the needs of the factory, in particular for a smooth flow of total orders even as the mix of orders fluctuates, it's possible to make the factory work better.

Moreover, the Japanese system helps fine-tune new products and catch embarrassing or dangerous errors before massive—and highly visible—public recalls are needed.

Finally, the lean selling system instills channel loyalty in the buyer and makes it extraordinarily hard for new competitors to gain share. This is a key reason Western mass-producers have had such a hard time making headway in the Japanese market. It is only in the past few years, as Western firms such as BMW and Daimler-Benz have made the necessary investment in their own distribution channels, that import sales have become significant, rising to 5 percent in 1990 compared with less than 1 percent for decades.[15]

## Information Technology and Lean Customer Relations

As we've pointed out, the Japanese companies are well aware of their selling costs, particularly for door-to-door sales, just as they are aware of their costs in every other area of production. They believe that the most promising way to cut back these costs lies in the area of information technology. To see how information technology might work, let's take one more trip to our Corolla dealer.

The first thing a consumer encounters on entering a Corolla dealership today is an elaborate computer display. Each Corolla owner in the Corolla family has a membership card that can be inserted in the display, just as one would insert a card in a bank machine. The display then shows all the system's information on that buyer's household and asks if anything has changed. If it has, the machine invites the owner to enter new information. The system then makes a suggestion about the models most appropriate to the household's needs, including current prices. A sample of each model is usually on display in the showroom immediately adjacent to the computer display.

At this point, if the owner is seriously interested in buying, he or she can approach the sales desk where the seven or eight team members are seated and discuss the particulars of a sale. Cars sold in this manner are rising steadily in Japan (from about 20 percent at present), and the companies hope that over the long term, they can largely deal with most existing owners in this way. Indeed, at some point in the future, they hope that the same information will be available at every owner's home on a computer or television screen.

The customer has access to other data bases as well—on everything from obtaining financing and insurance to parking permits (necessary in many Japanese cities before you can buy a car). Customers can also access information on secondhand cars should they want to buy one and obtain data on all their transactions with the dealer for service and inspections.

While each owner will still know an individual in the sales network to contact in case of difficulty, most of the sales force can now be directed to "conquest" sales of owners who are currently loyal to other brands. The end result, the companies hope, should be that the selling cost for the average new car will drop substantially, but the information harvested from consumers and the feeling of channel loyalty will be retained. If the Japanese producers can accomplish this goal and then transfer this truly lean selling system across the world, the system of lean production will be complete.

What we see in Japan, then, is that distribution is a fully integrated part of the entire production system. It is not simply an expensive door-to-door selling system. In essence it is a system that provides a high level of service to the customer and a high level of real feedback to the manufacturer. When product planning, marketing and distribution costs, together with the benefits of more accurate matching of production to demand (meaning less discounting and distress selling) and better production scheduling (meaning a more efficient factory) are added in, the Japanese system is already delivering a higher level of service for much lower real cost than Western analysts have realized. When information technology is fully added to the system to produce a truly lean distribution network, it should be possible to eliminate yet another of the inherent compromises in mass production. Just as quality costs less, not more, in the lean factory and as designing products faster reduces costs and errors, selling cars in a lean fashion with a high level of service should be possible at much lower real costs than in mass production with its low level of service. Lean distribution will form the front end of a system that is driven by the needs of the customer, not by the needs of the factory. In an increasingly competitive world market where, as we saw in Chapter 5, more affluent customers are seeking—and able to pay for—a greater choice in personal transportation, this reorientation of the entire mass-production system will be critical for survival.

Today there is much discussion in the West of the inadequacies of the distribution system. Customers are unhappy, the assemblers are unhappy, and the dealers are only marginally profitable. However, discussion in the West of the future of automotive distribution has so far focused on finding a new winning format for dealers—megadealers, publicly owned dealer chains, separate sales and service outlets, or lifetime-care schemes that keep the customer coming back for repairs for more than the first three years of the product's life. As we've seen, however, this isn't the right way of looking at the issue. Instead, we must think of distribution in a wider context, as an integral part of a customer-focused lean-production system. The winning formats that fit this system may turn out to be quite different from our current expectations. Indeed, we may end up with not just one winning format but several to suit different types of customers, products, and market segments.

With this review of distribution, we've now looked at all the steps in the immensely complex task of producing a motor vehicle. One of the features of lean production we noted in every chapter was the need for close coordination between the many steps, often involving face-to-face contact. Surprisingly, this is true even of distribution, where a truly lean distribution system probably requires a production system in or very near the market of sale.

Because of the great distances between the world's major markets and the persistence of trade barriers, this suggests in turn that lean producers wishing to succeed in the global motor industry over the long term will need to develop complete production/distribution systems in each major region. But how does a company create and manage such a global network of production complexes? This is the challenge we'll examine in the next chapter.

# MANAGING THE LEAN ENTERPRISE

## 8

The steps of production, from the day a new automobile design is initiated to the day an owner drives the car away, are only part of the total production process. For these steps to succeed, money must be available to underwrite the multiyear development effort, a highly trained and motivated staff must be in place, and activities occurring at different places around the world must be coordinated. Although no company, so far, has succeeded fully in doing so, we believe that lean producers must approach the tasks of finance, personnel management, and global coordination in a very different way from mass-producers. Collectively, the lean approach to these activities, if it can be perfected, will complete the lean enterprise.

### FINANCE

Henry Ford, as you may remember from Chapter 2, needed no external financing. By selling cars for cash faster than his suppliers came to collect their money, he managed to remain completely self-sufficient, while running a giant business entirely owned by his immediate family. Indeed, when Henry Ford II took over from his grandfather in 1945, he inquired where the company's financial reserves were on deposit; he learned that the money, about $700 million, was all kept in cash in the company vault.[1] The first Henry Ford had never even deposited money in the bank, much less borrowed from one, and all the company's stock at the time of his death was held by family members.

Hardly anyone else in the mass-production auto industry was ever so independent. Most of the major car companies in the West were publicly held from an early date in their histories, as the financial demands of rapid growth caused the founders to convert from family or private financing to the stock market. (The Ford Motor Company finally traveled down this path in 1956.) Of course, in many cases (Peugeot, Fiat, Ford) the founding family retained and retains a dominant share of the stock.

After World War II, a number of companies in Europe found a new source of financing in the form of public ownership. A new Volkswagen was established with the German government as its major stockholder, while Renault, Alfa Romeo, the Spanish company Seat (which is now owned by Volkswagen), and British Leyland came under public control at different times for different reasons. In the case of Renault, the French state saw the company as an engine of growth that would introduce mass-production principles to the entire country. The Italian, Spanish, and British governments, by contrast, were unwilling to see one of their largest companies fail. In any event, this era of public ownership has now largely come to an end. With the privatization of Rover, Alfa Romeo, and Seat in the late 1980s, and the recent sale of the German federal government's stake in Volkswagen, only Renault remains under public control. Its senior managers have vigorously argued that it, too, should be privatized.

Thus, shares of practically all the Western auto companies are traded on public exchanges. So are the Japanese lean producers, but there the similarities end. You may remember from Chapter 3 the story of Toyota's efforts to get on a sound financial footing after World War II. Its experience can be extended to the other Japanese auto makers.

In the first era of Japanese industrialization, after the Meiji Restoration in 1870, large companies were financed through the *zaibatsu*. These family-owned holding companies controlled industrial empires that consisted of a large company in each of the major sectors—steel, shipbuilding, construction, insurance, finance. Each *zaibatsu* included a bank, and the deposits in the bank were the major source of funds for investments by the companies in the group.

The Americans eliminated these tightly organized groupings during their post–World War II occupation of Japan. After the Americans left, the *zaibatsu* were replaced by a new form of industrial finance, the *keiretsu*. Each *keiretsu* consists of perhaps twenty major companies, one in each industrial sector. Unlike the *zaibatsu*, there is no holding company at the top of the organization. Nor are the companies legally united. Rather, they're held together by cross-locking equity structures—each company owns a portion of every other company's equity in a circular pattern—and a sense of reciprocal obligation. Toyota, for example, is affiliated with the Mitsui *keiretsu*, while Mazda is a member of Sumitomo, and Mitsubishi Motor Company is a member of Mitsubishi. Among the key companies in every group are a bank, an insurance company, and a trading company. Each of these has substantial cash resources that can be made available to the members of the group. In fact, their key purpose is to help each other raise investment funds.

These groupings arose only gradually as Japan rebuilt after the Americans left. The equity in the previous *zaibatsu* had been declared void in 1945. The Japanese companies initially were financed almost entirely by loans supplied by the big Tokyo banks and guaranteed by the American government. Since companies had only these loans and their physical assets, their equity was very modest. As the economy took off and many companies became profitable, they began to worry about being bought up by foreigners. They also distrusted the arm's-length stock market as the primary means of generating equity, because they couldn't imagine a system in which there was no reciprocal obligation.

To address these concerns, the growing companies of the 1950s and 1960s hit upon the idea of selling equity to each other, often with no cash changing hands. So each member of the prewar groups, and some newcomers as well, joined the new *keiretsu* in which the equity went around in a circle.

The large groups were essentially privately held—but on a massive scale. That is, their stocks traded in small volumes on the highly volatile Tokyo stock market, but the stock that really counted was never for sale. The Americans and other foreigners discovered this fact after 1971. That year equity in Japan was liberalized to permit foreign majority control of any company, but none of the *keiretsu* members was willing to sell its "captive" shares at any price. So few companies actually could be bought.

The system was glued together partly by a sense of reciprocal obligation—each member of the group held every other member's stock in a sort of trust. However, if the sense of obligation faltered, the more practical factor of hostage equity kept selling in check: If one company considered selling its stake in another to an outsider seeking to gain control, the second company could retaliate by selling the first company's equity to outsiders as well. No one sold.

A variant of this system was soon extended downward into the supplier groups, as well. We saw in Chapter 3 how Toyota spun off supplier companies, such as Nippondenso and Toyoda Gosei. Toyota held an equity stake in these companies, and they had a small equity stake in Toyota. Soon the Toyota industrial group exhibited some of the circular equity structure seen in the *keiretsu*, although Toyota held a strong position at the center.

The recent attempts of the American raider T. Boone Pickens to seize control of Koito, a Toyota group member, show just how powerful the group system is. Toyota owns only about 15 percent of Koito, and Pickens was able to buy up shares totaling more than 26 percent. Yet, he couldn't gain a seat on the Koito board. In addition, no other shares seemed to be for sale, even at an offering price well above what the shares would fetch on the open market.

This system of group equity has been exasperating to Western companies and governments because its logic is so different. Japanese companies, with what at first appears to be a public equity structure, are in reality privately held. This arrangement would not be allowed under the investment laws of the United States and a number of European countries—companies would have to explain that only some of their stock was actually for sale. While we believe the *keiretsu* and industrial groups are in fact the most dynamic and efficient system of industrial finance yet devised, they are not adequately understood in the West.

Besides providing members protection against hostile takeovers, an often cited advantage of the *keiretsu* system is the low cost of funds for group members. The inexpensive funds come in two forms.[2] First, many Japanese companies pay hardly any dividends. Typically, they pay a 10-percent yield on the par value of their stock, where the par value, established at the time of the initial equity trades in the 1950s, is substantially zero. So Toyota stock, for example, in fiscal 1989 paid a dividend of 18.5 yen or 10 percent of earnings, while Nissan paid only 7 yen or 7 percent of earnings.

Second, in the 1980s, the booming Tokyo stock market permitted the Japanese auto companies to issue vast amounts of new stock in the form of convertible bonds that could be converted to stock if a company's stock reached a certain price in the market. Buyers of these bonds were willing therefore to accept very low interest rates on the presumption that their real return would come from the stock conversion in the continually advancing Tokyo market. During the 1980s Toyota issued $6.2 billion in convertible bonds at interest rates from 1.2 to 4 percent, a cost of capital far below that available to Western auto companies. Even much weaker companies such as Isuzu and Fuji Heavy Industries (Subaru) were able to obtain low-cost financing through this means.[2]

How long this second form of low-cost fund raising can continue is an interesting question. On the one hand, the slump in the Tokyo market in 1990 suddenly made investors aware that conversion will not always be possible, and the issuance of convertible bonds stopped, at least temporarily. On the other hand, Japan is still a country of obsessive savers, and these savings must find some outlet.

However, even in the absence of cheap investment funds, the Japanese group system still confers a significant competitive advantage, specifically in ensuring that investment funds are wisely spent. For proof of this hypothesis, we need look no farther than the way financial systems in Japan and the West deal with companies in distress. The key Japanese instance is the Mazda turnaround in 1974. Up to that time, Mazda had been run by its founding family, which was strongly oriented to product engineering. The hallmark of the company was its fuel-hungry but technically advanced Wankel rotary engine. When energy prices suddenly soared in 1973, Mazda faced a major problem. It needed an entirely new set of fuel-efficient piston engines, and it needed a new range of models for the new engines.

The company faced another pressing problem as well. Mazda had been charging higher-than-average prices for cars in its market segment, even though the Wankel was cheaper to make than ordinary engines. The reason for the high prices was Mazda's inefficient production system, which resembled mass production much more than lean. Up to this point, Mazda's cars could command premium prices because of the high-tech image of the Wankel motor. Scrapping the Wankel meant Mazda would now be selling ordinary cars, so prices had to come down. And for this drop to occur, it was essential that Mazda reform its production system.

Mazda's salvation came from the Sumitomo group, which controlled the car company's equity through cross links. The Sumitomo bank sent a team of executives to replace the family management. These new executives' key decision was to copy the Toyota Production System at Mazda's Hiroshima production complex, so that Mazda could become cost- and quality-competitive with the best Japanese companies. A second key decision was to provide massive loans for new engines and a new model range, so Mazda could expand, rather than contract, its market presence.

The contrast with British and American practice in the 1970s and 1980s is striking. When British Leyland and Chrysler began to founder, their bankers and institutional investors, numbering in the hundreds, were mainly concerned with how to minimize their exposure. In both cases, the stock was widely held and no effective organization of stockholders existed to voice their concerns. The outside members of the company boards neither understood the real problems nor knew what to do. Instead, the boards remained passive, the banks wrote off their loans, and the institutional investors simply sold their stock at a loss and walked away.

British Leyland ended up under direct government control for a decade, while Chrysler needed a government-guaranteed loan to regain its momentum. However, neither firm inspired enough confidence in governments or investors to qualify for more than the minimal amount of assistance to keep going. With tight constraints on product-development funding, both companies struggled through the 1980s. More significant, in neither case was the financial system or the government able to tackle the real problem: dysfunctional mass-production systems that could no longer compete in the world market.

The investment finance systems behind the other Western producers in Europe have been more effective, at least in providing companies the funds necessary to weather crises. This is because a single large shareholder with a long-term outlook has controlled the destiny of most European companies: the Agnelli family at Fiat, the Peugeot and Michelin families at PSA, the

Quandt family at BMW, the Handelsbank at Volvo, the Wallenberg family at Saab, the Porsche/Peich family at Porsche, and the Deutsche Bank at Mercedes. Renault, of course, is still state-owned, and Volkswagen had a large government shareholding until recently. Thus no company has been "friendless," with no significant shareholder committed to the company and no strong relation to a major bank.

Nevertheless, although the Japanese groups do make mistakes, in some instances very big ones, the *keiretsu* system on average has exhibited superior performance compared with both the Anglo-Saxon (American and British) and continental European systems of finance. Western finance tends to be either impatient and largely uninformed about a company's problems (as in the case of the American and British institutional investors and banks that dump their stock and loans at the first sign of trouble) or patient but passive (as in the case of outside directors in the United States and Britain and the family shareholders in continental Europe). The latter have often failed to confront the problem of clear slippage in competitive position until very late in the game.

By contrast, the Japanese group system is patient and extremely long-term in orientation—but very well informed and highly critical of inadequate performance. The groups can afford to invest heavily to finance corporate turnarounds, because their considerable knowledge reduces the risks of failure.

## CAREER LADDERS

As we've noted at a number of points, mass production provides no career progression for production workers. Engineers, financial analysts, and marketing specialists progress through technical expertise. The progression for the general manager is through ever higher levels of the corporate hierarchy. All three paths are dysfunctional for the organization as a whole. The lean enterprise, by contrast, strives to provide every employee with a clear career path, although these are very different from those of mass production.

To begin with, every employee begins by working on the production line for some period of time. For example, while visiting the Honda plant in Marysville, Ohio, recently, we asked to meet with the external affairs director, the person at Honda who handles relations with governments and the general public. He was unavailable, we were told: He had just joined the company and was busy assembling cars. The best lean producers believe that the point of production is where value is truly added, not through indirect managerial activities, and that all employees need to understand this fact as soon as they enter the company.

Those who stay in the factory grow increasingly able to solve problems. Management stresses that problem-solving is the most important part of any job. Management's objective is to give employees increasingly challenging problems to solve in order to test continually their skills, even when, unlike in Western companies, no promotion up a ladder to section head, say, or factory manager is possible. Higher pay comes largely on the basis of seniority, with performance bonuses as well. In other words, the lean manufacturers, which operate without much of the hierarchy we find in Western companies, try to make employees understand that their capacity to solve increasingly difficult problems is the most meaningful type of advancement they can achieve, even if their titles don't change.

For those employees with a specialized skill—mechanical engineering is the most common—the lean producer attempts to harness the skill to a team process where it will be of maximum use. We saw how this technique works in Chapter 5. We also saw how team members are shifted to subsequent teams and how they may be asked to learn entirely new skills as they move through their careers.

For employees who are needed for general management, the contrast between mass production and lean is equally striking. Because decision-making and problem-solving are pushed far down the ladder in the lean company, it has much less need for layers of middle and senior managers to send orders down the hierarchy and transfer information back up. Instead, the key functions for managers are to tie the supplier organizations to the assembler organization and to tie together geographically dispersed units of the company. Typically, the company sends managers at the midcareer level to high-level positions in the supplier companies in the assembler's group and rotates mid- and senior-level managers between the company's operations, particularly the foreign operations.

These practices have two advantages. They create a complex network of interpersonal relations, so the assembler and the suppliers and those in the company's international operations know each other through personal contacts. They are also the conduit through which the company's culture is spread into the supplier system and to new regions.

## GEOGRAPHIC SPREAD

The world at large, including, we must note, a number of the Japanese lean producers, does not yet understand a vital feature of lean production. This mode of production achieves its highest efficiency, quality, and flexibility when all activities from design to assembly occur in the same place. As a senior Honda executive recently remarked, "We wish we could design, engineer, fabricate, and assemble the entire car in one large room, so that everyone involved could be in face-to-face contact with everyone else." And, as we saw in the last chapter, the final step in the system, lean sales and service, cannot work at all without a production system located in the same area as the sales market.

For this reason, lean producers in the 1990s will need to create top-to-bottom, paper-concept to finished-car manufacturing systems in the three great markets of the world—North America, Europe, and East Asia (centered on Japan). This process is farthest along in North America, where the Japanese firms began opening assembly plants in 1982. Eleven were in operation by the end of the decade and in 1990 accounted for slightly more than 20 percent of automobile assemblies in North America, as shown in figures 8.1 and 8.2.

Doing the whole job in one large room will not be possible, of course. Nor will it be possible to do it even in an area as constricted as Toyota City, but the geographic pattern of lean production in North America is already clear. The transplant assembly operations (with the exception of NUMMI) are located within a 300-mile radius in the American-Canadian Midwest. The vehicles assembled in these plants at first contained only about 20 percent U.S. and Canadian content, but this figure rose steadily to about 60 percent in 1990, and we expect it will reach 75 percent by the late 1990s.

The supplier plants, some old and some new, are mostly located nearby, so parts can be shipped from supplier to assembler in less than a day's drive. (Comparisons with Japan in terms

**FIGURE 8.1**

**Japanese Transplant Share of North American Automobile Production, 1982–1990**

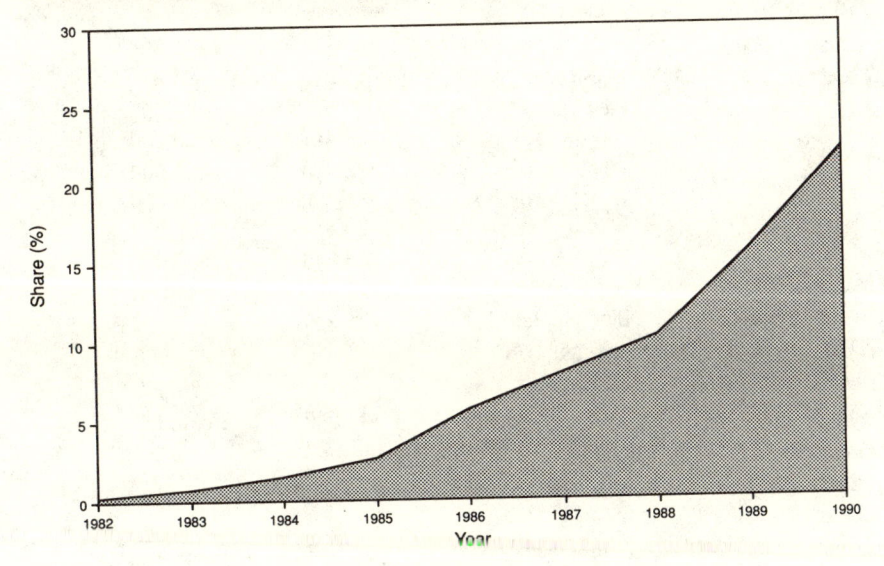

*Note:* 1990 is estimated based on production in the first three months.
*Source:* Calculated by the authors from *Ward's Automotive Reports*

of the geographic concentration of suppliers can be quite misleading. Road congestion in Japan is so severe that suppliers located within 50 kilometers [30 miles] of an assembly plant may actually need more time to deliver their parts than suppliers 200 kilometers [125 miles] from the Japanese assembly plants located in rural areas in the U.S.-Canadian Midwest.)

Honda, Toyota, Nissan, Mazda, and Mitsubishi have now all established North American product- and process-engineering operations as well. Honda, true to its conviction about doing it all in one place, has located its engineering center at its Marysville, Ohio, complex, while the other companies have all located in the Detroit area. Their reasons: They want to be close to the headquarters of U.S. suppliers and be able to recruit engineers easily in Detroit.

These centers are growing rapidly, although it will be past the turn of the century before they approach the size of the Big Three American companies in Detroit. However, they are already

### FIGURE 8.2

**Japanese Transplant Production Facilities in North America**

| Firm | Location | 1989 Production | Announced Capacity | Notes |
|---|---|---|---|---|
| *Assembly Plants:* | | | | |
| Honda | Marysville, OH | 351,670 | 360,000 | |
| | East Liberty, OH | | 150,000 | (1) |
| | Alliston, ON | 86,447 | 100,000 | |
| NUMMI | Fremont, CA | 192,235 | 340,000 | (2) |
| Toyota | Georgetown, KY | 151,150 | 240,000 | |
| | Cambridge, ON | 20,859 | 50,000 | |
| Nissan | Smyrna, TN | 238,640 | 480,000 | (3) |
| Mazda | Flat Rock, MI | 216,200 | 240,000 | |
| Diamond Star | Bloomington, IL | 91,839 | 240,000 | (4) |
| CAMI | Ingersoll, ON | | 200,000 | (5) |
| SIA | Lafayette, IN | | 120,000 | (6) |
| ASSEMBLY TOTAL | | 1,349,000 | 2,520,000 | |
| *Engine Plants:* | | | | |
| Honda | Anna, OH | | | |
| Nissan | Smyrna, TN | | | |
| Toyota | Georgetown, KY | | | |
| TOTAL ENGINES | | | | |

*Notes:* (1) Commenced operations in 1989.
(2) General Motors/Toyota joint venture. Truck assembly line being added.
(3) Second assembly line being added.
(4) Chrysler/Mitsubishi joint venture.
(5) General Motors/Suzuki joint venture.
(6) Subaru/Isuzu joint venture.

Announced capacity is typically for two eight-hour shifts per day, five days per week. Thus use of overtime can permit production in excess of "capacity" up to about 120 percent of capacity for extended periods.

*Source:* 1989 production from *Ward's Automotive Reports.* Capacity plans from company announcements.

---

doing significant design and engineering. The body alterations needed to create the Honda Accord coupe and station wagon from the initial Honda Accord sedan were engineered at Marysville, and all the production dies were cut there as well. The coupe and station wagon will be assembled exclusively at Marysville for the whole world, with exports to Japan and Europe. Nissan's Ann Arbor, Michigan, engineering center is doing similar engineering

# MANAGING THE LEAN ENTERPRISE

work on the coupe version of its new Sentra model, which will be assembled exclusively at Smyrna, Tennessee, for all world markets.

In Europe, the Japanese advance has been much slower, for reasons that we'll examine in the next chapter. However, the pace of Japanese investment is now rapidly building, as shown in Figure 8.3. We anticipate that by the end of the 1990s, several Japanese assembler firms will have top-to-bottom production systems in place in Europe as well.

**FIGURE 8.3**

## Japanese Transplant Production Facilities in Europe

| Firm | Location | 1988 Production | Announced Mid-1990s Capacity | Potential Additional Capacity |
|---|---|---|---|---|
| *Assembly Plants:* | | | | |
| Nissan | Washington, UK | 57,000 | 200,000 | 200,000 |
|  | Barcelona, Spain | 76,000 | 150,000 | |
| Honda | Swindon, UK | | 110,000 | 260,000 |
|  | Longbridge, UK | 4,000 | 40,000 | 400,000 (1) |
| Toyota | Burnaston, UK | | 200,000 | 200,000 |
|  | Hannover, Germany | | 15,000 | (2) |
|  | Lisbon, Portugal | 14,000 | 15,000 | |
| Isuzu | Luton, UK | 35,000 | 80,000 | (3) |
| Suzuki | Linares, Spain | 22,000 | 50,000 | (4) |
|  | Esztergom, Hungary | | 50,000 | |
| Mazda | ? | | | 100,000 (5) |
| Mitsubishi | ? | | | 100,000 (5) |
| TOTAL ASSEMBLY | | 208,000 | 940,000 | 1,260,000 |
| *Engine Plants:* | | | | |
| Nissan | Washington, UK | | 200,000 | 200,000 |
| Honda | Swindon, UK | | 70,000 | 330,000 |
| Toyota | Shotton, UK | | 200,000 | 200,000 |
| TOTAL ENGINES | | | 470,000 | 730,000 |

Notes: (1) Production by Rover for Honda. Potential figure supposes a takeover of Rover by Honda.
(2) Toyota vehicles assembled by Volkswagen.
(3) Joint venture with General Motors.
(4) Excluding the assembly of Land Rovers.
(5) New plants under discussion.

Source: Comité des Constructeurs Français d'Automobiles, *Reportoire Mondial*, Paris, December 1989, p. 9, elaborated by the authors

## THE ADVANTAGES OF GLOBAL ENTERPRISE

Besides the vital advantage of doing it all in one place near the point of sale, creating a top-to-bottom manufacturing system in each of the world's major markets benefits a company in five ways, compared with rivals trying to manufacture and export from a single region.

First, and most obvious, it provides protection from trade barriers and currency shifts. For the company producing at one site in one region, such as Jaguar in Britain or Saab in Sweden, currency shifts can produce an export windfall—for example, the high profits these companies obtained in the United States in the mid-1980s, when the dollar was strong in relation to European currencies.

But disaster is equally likely. Between 1987 and 1989, Jaguar and Saab didn't get worse at making cars. In fact, our IMVP assembly plant survey showed a modest improvement in manufacturing productivity and product quality. Moreover, both companies introduced new models that strengthened their product range. Yet during this period, a weaker currency in the United States, each company's primary export market, converted Jaguar and Saab from high flyers to near-bankrupts. They were then absorbed, respectively, by Ford and General Motors, companies with multiregional production bases.

For large companies that want to capture a substantial fraction of each regional market, the lesson of the 1980s is clear: There is simply no substitute for within-the-region production. The purchase of cars and trucks accounts for about 15 percent of personal consumption in North America, Europe, and Japan. This is such a large number—$240 billion per year in the case of North America—that it's difficult to imagine offsetting exports that could balance trade among regions when one (Japan) produces massive numbers of motor vehicles and the other regions consume them.

The experience of the 1980s also suggests that if trade barriers do not arise to rebalance motor-vehicle trade, currency shifts will. These methods do have different consequences. Government-imposed quotas on the import of finished units tend to make importers rich as they raise prices to ration demand, while currency shifts do the opposite. However, in either case, the fact remains that in the long run producers must either locate within the market of sale (as the Japanese are doing in North America and Europe) or cede that portion of the world motor-vehicle market (as the European volume producers seem to be doing in North America).

A second advantage for the company developing a multi-regional, top-to-bottom production system is rich product diversity. As we saw in Chapter 5, the motor-vehicle market in Europe, North America, and Japan is progressively fragmenting, with no end in sight. A lean production system can gain most economies of scale at much lower volume per individual product compared with mass production, as we also saw in Chapter 5. However, achieving this goal presumes that the variety of products can be assembled in sequence on one large production line using several sizes of engine and transmission from a large engine plant and a large transmission plant. So companies with higher production volumes for all their products combined still have a competitive advantage. As long as corporate management can deal with the complexity, being big still means being better, and being big in the 1990s requires that companies manufacture in each of the major regions.

Equally important, consumers in the three regions continue to demand different types of products and—this point is key—to attach different images to the same product. Consider the example of German luxury cars sold as taxis in Germany to create a volume base for their manufacturers, but sold in North America and Japan at much lower volumes and much higher prices as luxury goods. Similarly, Honda has recently pocketed healthy profits by exporting its Accord coupe, built and sold in the United States at high volume, as a much more luxurious, limited-volume product for the Japanese market.

Honda seems to have been the first to see the advantage of this approach. It plans during the 1990s to develop a set of products unique to each major region. These will be produced within the region to serve volume segments in that region. The company will then export these products to other regions to fill market niches where it hopes their limited volume and exclusivity will permit charging higher prices.

If this approach is carried to its logical conclusion, the multi-regional producer will have an intracompany product portfolio and trade flow as shown in Figure 8.4, where the majority of

#### FIGURE 8.4

**Cross-Regional Product Flows Within a Company**

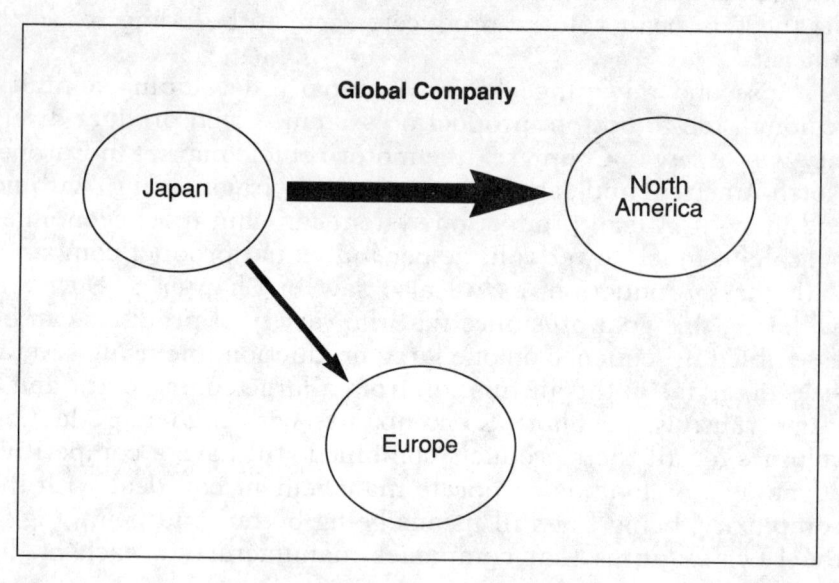

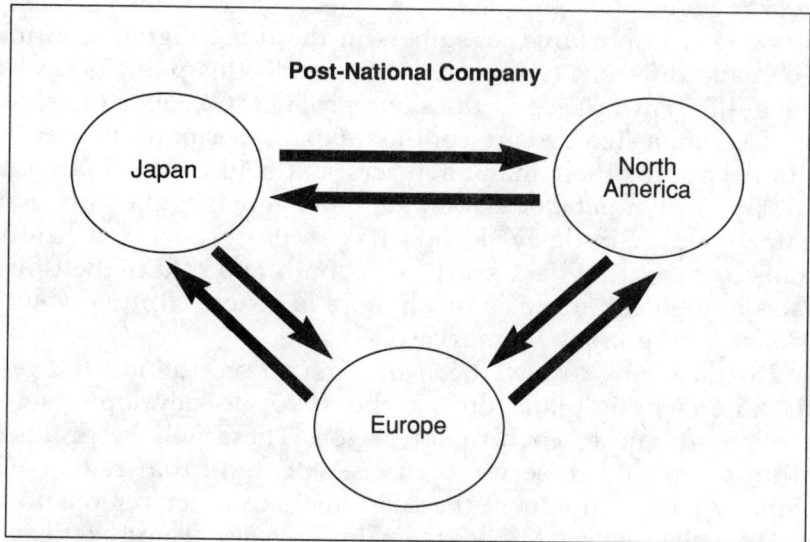

**Global Company**

Japan → North America
Japan → Europe

**Post-National Company**

Japan ⇄ North America
Japan ⇄ Europe
Europe ⇄ North America

demand is met by the production system within each region and cross-regional trade is reasonably balanced.

A third advantage the multiregional producer can achieve over the single-region producer is the sophistication managers gain through exposure to many different environments. Sophistication is subjective, of course, but in our interactions across the world with executives in all the major assembler and supplier firms, we've been struck by how much extra perspective managers gain from trying to manufacture products in different environments.

For example, we are convinced that one reason Ford has performed better than General Motors in recent years is simply that Ford has more of its production activities outside the United States and moves more personnel back and forth between its different international operating units. It's now rare to meet a senior executive at Ford who hasn't spent years managing operations outside the United States.

By contrast, while GM has many foreign subsidiaries, it is still common to meet GM executives who have punched their ticket with a two-year tour at Opel in Germany or at GM of Europe in Switzerland, but who otherwise haven't worked outside the American Midwest. The wider exposure at Ford produces a higher level of sophistication in operations management. Since managers have been exposed to radically different ways of solving problems, they also have the flexibility to think more creatively about strategic issues facing the company. (Gaining the full benefit of international operations obviously requires a sophisticated personnel system to rotate managers in the most productive ways, an issue we'll return to shortly.)

A fourth advantage for the multiregional producer is protection against the regional cyclicality of the motor-vehicle market. Motor vehicles are the foremost among what economists call durable goods. With some patching, owners can almost always get their cars to run a bit farther. So motor-vehicle sales in every country tend to be more volatile than the general economy. However, the world major markets don't go up and down at exactly the same time; for example, the Japanese market was booming at the close of the eighties, while the American market had gone soft. Thus, a company with a presence in all major markets has more protection against cyclicality.

Establishing a global production system is particularly important for those U.S. companies relying predominantly on the especially cyclical North American market. The Japanese companies still sell the majority of their cars in the Japanese market, which, for reasons we'll examine in the next chapter, is much less cyclical. They'll therefore find it easier to plow through the next automotive recession in North America and will cut prices if necessary to sustain smooth production at their new transplant facilities. By contrast, General Motors and Chrysler largely operate and sell in the United States and Canada. Any fall-off in sales will force them to take money away from product-development activities and their foreign alliances to cover short-term operating costs.

The product-development consequences will become apparent only in the mid-1990s, when these U.S. companies will probably suffer further share losses. However, the slow market conditions of 1989 and 1990 already have had some effects. Chrysler reduced its share in Mitsubishi Motors from 24 to 12 percent and GM reduced its equity stake in Isuzu from 44 to 38 percent. These actions to raise cash are moving these companies in precisely the wrong direction in terms of establishing a global production presence.

This fact becomes apparent when we consider the final advantage of developing a full-fledged production system in each of the major markets: Doing so denies competitors defended markets from which to skim profits to use in competitive battles elsewhere in the world.

The Japanese domestic market during the 1980s provides the most striking example of what can happen when foreign companies cede a major regional market to domestic companies. The Western companies could have pushed hard to buy the weaker Japanese companies—Isuzu and Suzuki, in GM's case, and perhaps Mazda as well (Ford). This would no doubt have led to "investment friction" in Japan, where, as we have just seen, the group equity structure effectively excludes foreigners unless the group consciously decides to include them. However, this is an issue that will have to be faced shortly in any case, and it would have been in the interest of the Western firms to push very hard on this point.

Instead, they pushed for trade liberalization, so they could more easily export finished cars and parts to Japan. Liberalizing trade was an uphill struggle—even in the absence of trade barriers—because the Americans really had nothing to sell that was competitive, either on price or quality, in the Japanese auto market except a few novelty products, such as Cadillac limousines, the car of preference for Japanese gangsters until they transferred their loyalties to Mercedes in the 1980s.

Meanwhile, the Japanese companies received a windfall from North American and European quotas. When the Japanese were told they could sell only a fraction of the cars they had sold previously, they simply raised their prices until sales fell to the required level. And they reaped huge profits in the process. Indeed, the Western quotas are arguably the biggest public-policy boost the Japanese auto industry has ever received—more useful than the Ministry of International Trade and Industry (MITI) ever was in Japan. The Japanese companies used their profits to wage a market-share war in Japan, probably selling below cost in many cases and ensuring that Western importers would have little success selling there, even if there were no trade barriers at all.

Then, in the late 1980s, the situation reversed. The Japanese companies used large profits from the booming domestic market (where the Japanese government did help by sharply reducing the commodity tax on car purchases) to underwrite their massive investments in production facilities in North America and Europe. They could move ahead without fear of U.S. producers filing suit that they were dumping or practising other trade retaliation, because their profits were not used to sell cars abroad below Japanese prices. Instead, their profits were used for capital investments and new products, such as the Toyota Lexus LS400 and the Nissan Infiniti Q45, designed primarily for the U.S. and European markets.

The U.S. and European companies, with no production presence in Japan, made some profits themselves in the strong Japanese market through a trickle of imports but missed most of the opportunity. The failure to establish a manufacturing presence in Japan or elsewhere in East Asia—to seriously challenge Toyota, Nissan, and Honda in their home market and take away this rich profit lode—is surely one of the West's worst competitive lapses.

## MANAGING THE GLOBAL ENTERPRISE

Given the overwhelming evidence that a multiregional production presence is now essential for success in the motor-vehicle industry, one question remains: how to manage a lean, global enterprise consisting of three top-to-bottom production complexes in the 1990s, and perhaps several more in the twenty-first century (for instance, in India for the southern Asia market, in Brazil and Argentina for the Latin American market, in Indonesia or Australia for the Oceanic market, and even in South Africa—if the present movement toward rejoining the world community continues—for the southern Africa region).

This is not a trivial management issue. Indeed, dynamic and effective management of global production organizations has largely defied the ingenuity of the automotive mass-producers over nearly a century of trying.

The first auto company to pursue a global manufacturing strategy was Ford.[3] The present Ford Motor Company was founded in 1903 to manufacture the original Model A. By 1905, even though annual production still totaled less than a thousand units, Henry Ford had established a Canadian manufacturing plant to assemble Ford cars for sale in Canada. In 1911, three years after the introduction of the Model T, Ford opened an assembly plant at Manchester in England. By 1926, Ford was operating assembly plants in nineteen countries.

However, these steps hardly constituted serious internationalization. Ford's primary motivations were to reduce shipping costs—parts were cheaper to ship than finished units—and to surmount tariffs. Then as now, these were usually higher on finished units than on parts. Henry Ford made clear that all design and as much component fabrication as possible were to be retained in Detroit. In addition, the foreign branch plants, as they were called, were almost always under the management of Americans sent from Detroit.

This pattern continued throughout the 1920s. However, as one country after another erected trade barriers after the collapse of the world economy in 1929, Ford was forced to go farther. He built a fully integrated manufacturing complex at Dagenham in England in 1931 and a similar, though smaller complex, at Cologne in the same year. By the mid-1930s, these plants were producing practically all the parts for Ford's products. Much more radical from Henry Ford's perspective, they manufactured a new product, the Model Y, not produced in the United States. This was Henry Ford's belated acknowledgment that Europeans did not want to drive American-style large cars.

We must remember, however, that the Model Y was designed in Detroit, and many of the tools for its manufacture were made there as well. While English engineers suggested ways to accommodate the car to European tastes, the Model Y and all Ford products of the 1930s were practically 100-percent American engineered.

Only after the war did Ford of England and Ford of Germany begin to hire their own product-development engineers, and not until 1961, with the introduction of the Ford Anglia, was a Ford product for the first time completely designed in a foreign country.[4] This development occurred exactly fifty years after Ford began its European assembly operations at Trafford Park, Manchester.

By this point, the Ford Motor Company had turned 180 degrees from its original practice. Where Henry Ford had demanded 100-percent control of the product and ensured that all manufacturing decisions came from Detroit, Henry Ford II presided over a remarkable process of decentralization in which the newly formed Ford of Europe shared no common products with Detroit. It also had limited personnel transfers—that is, only a few Americans in senior positions. It was in many ways a totally separate company in all respects save financial.

Because it recognized the emergence of a unified Western Europe before West German, French, or British companies did—becoming the first "European" company in Europe—Ford of Europe (established in 1967) became remarkably successful and was a key contributor to the survival of Ford in North America. Massive loans from Ford of Europe tided Ford over during the great North American auto depression of 1980 to 1982.

However, from the perspective of senior management in Detroit, the evolution of a largely decentralized company was far from ideal. By the 1970s, the company in North America had developed a wide range of products smaller than the standard-size American car of the 1950s. Many of these cars were identical in overall dimensions to the products developed separately by Ford of Europe. It seemed only logical that global standardization of products in each size class would produce enormous savings in development costs and manufacturing economies.

Ford's first attempt to standardize on a global basis was the Escort, introduced in 1979. A world design team was designated to develop this car with contributions from all of Ford's global operating companies. However, in the process a curious thing happened: The Europeans from Ford of Europe and the Americans from North American Automotive Operations managed to specify change after change in this "world" car to accommodate, respectively, European and American tastes and manufacturing preferences. On launch day, the European and American Escorts, although practically indistinguishable in external appearance, shared only two parts—the ashtray and an instrument panel brace.

In 1979, Ford bought a 25-percent stake in Mazda in Japan. Because Mazda also makes a full range of products from small to large, it seemed logical to integrate some of Mazda's products into the worldwide Ford product-planning and development process.

For a start, Ford established its own distribution channel in Japan (Autorama) and began selling restyled Mazda 121, 323, and 626 models there with "Ford" badges. These models are also sold as Fords in many markets in Southeast Asia. A bit later Ford decided to import a restyled version of Mazda's small 121 design to the United States from Korea, where it is assembled by KIA, a small firm in which Ford and Mazda both hold a small equity stake. This model is sold under the name Ford Festiva.

By the time the Ford-Mazda link was fully established, it was too late to consider a joint design exercise for Taurus/Sable (launched in 1985), but joint design was carried through on the new Mazda 323 and Ford Escort (launched in 1989 in Japan and in 1990 in the United States). A similar cross-regional exercise involving Ford of Europe and Ford North America (called CDW 27) is now under way on the new Ford Tempo/Topaz for the North American market and the Ford Sierra replacement in Europe, due in 1991.

Ford calls the process of joint design, with the lead role assigned to either Mazda at Hiroshima (for 323/Escort) or Ford North America in Dearborn (for the next generation of large cars to replace the Taurus/Sable) or Ford of Europe in the United Kingdom and Germany (for Tempo/Sierra), "Centers of Responsibility." Senior executives in the company have advocated this approach as the only way to control spiraling development costs for new products at a time when a greater variety of cars and trucks is needed in every regional market.

However, thus far full implementation of Centers of Responsibility has eluded Ford's best efforts. Ford of Europe argued that the new 323/Escort was too small for Europe and has pushed ahead with its own design for launch at the same time. Similarly, in 1989 it introduced a new Fiesta model in the next smaller size class after rejecting use of the Mazda 121 design (which was judged too small). Finally, executives in Europe are resisting the inclusion of their large car (the Scorpio) in the Taurus/Sable replacement program on the ground that no single design can satisfy both American and European consumers in this class of car. What's more, Mazda, while happy to act as lead designer on the 323/Escort project, has continued to design its own models for other size and market classes—121, Miata, 626, and 929—and these models continue to compete directly with Ford products in the major regional markets.

It is sobering to realize that even with its limited progress in globalizing design and production, Ford is still the clear leader among all companies, including the Japanese, in establishing itself as a truly global organization with design and production facilities in the three major markets. By contrast, Chrysler has only a tiny manufacturing presence outside North America, consisting of an agreement with Steyr in Austria to assemble 30,000 Chrysler vans each year (beginning in 1991). General Motors has a strong presence in Europe and Brazil but continues to run these operations as decentralized, stand-alone companies that hardly talk to its North American operations. Finally, the European-owned companies have either never started the globalization process or, as we'll see in a moment, made only halting progress at a few locations in developing countries.

The Japanese, by contrast, are now showing an intention to globalize after strong initial reluctance and have had some initial success. However, they face huge hurdles in the coming decade, as we will see in a moment.

## THE EUROPEAN FAILURE TO GAIN A GLOBAL PRESENCE

The European industry now trails the Americans and the Japanese in globalization, as shown in Figure 8.5. When we consider the experience of the Europeans, a fundamental axiom emerges: It is impossible to establish lean production on a global basis when you have not mastered it at home. The case of Volkswagen provides a good illustration.

In 1974, Volkswagen established an American assembly plant at Westmoreland, Pennsylvania. Its objective was to establish a lower-cost American manufacturing base as the German mark

## FIGURE 8.5

**The Internationalization of Vehicle Assembly, 1988**
**(% of total vehicles built by place of final assembly)**

|  | Home Country | Local Region | Other Regions |
|---|---|---|---|
| Ford | 53 | 13 | 34 |
| General Motors | 65 | 10 | 25 |
| Volkswagen Group | 56 | 25 | 19 |
| Fiat (excl Iveco) | 79 | 11 | 10 |
| Renault (excl RVI) | 61 | 34 | 5 |
| PSA | 77 | 20 | 3 |
| Honda | 72 | 3 | 25 |
| Nissan | 75 | 4 | 21 |
| Mazda (incl Kia) | 65 | 20 | 15 |
| Toyota | 89 | 2 | 9 |
| Mitsubishi | 80 | 13 | 7 |

*Notes:* Excludes double counting of kits assembled abroad.

Local Regions: Americans—USA, Canada, and Mexico; Europeans—EEC, EFTA, Poland, Turkey, and Yugoslavia; Japanese—Japan, South Korea, Taiwan, Thailand, Malaysia, Indonesia, Philippines.

*Source:* Estimated by the authors from Comité des Constructeurs Français d'Automobiles, *Repertoire Mondial*, Paris, December 1989.

---

appreciated and the Japanese producers stepped up their sales offensive in North America. However, Volkswagen knew nothing about lean production and staffed its U.S. plant with old-line manufacturing managers lured away from General Motors.

The results were disastrous. For one, cost savings failed to materialize. Equally damaging, the product adjustments made for the American market caused quality to tumble while alienating buyers attracted to traditional German products. After fifteen years of frustration, Volkswagen departed for Mexico in 1989, hoping that low wages there would provide the basis for reestablishing its historic position in the entry-level U.S. market.

Renault suffered an even more costly disaster. It purchased American Motors in 1979 with the intention of obtaining a low-cost presence in North America. However, Renault also had no understanding of lean production and made little headway in revitalizing some of the worst mass-production plants in North America.

By 1987 Renault had had enough. It sold out for a fraction of its purchase cost to Chrysler, which is belatedly attempting to reform these old facilities, in the first instance by closing the worst, such as the Kenosha, Wisconsin, assembly plant built in 1905.

We can appreciate the full cost of Renault's setback in the United States when we consider that the company now has no manufacturing presence outside France, Spain, and Portugal, three of the most protected markets in Europe, excepting a sole assembly plant in Belgium. Volkswagen at least retains a half interest in the AutoLatina operation in Brazil, owns an integrated manufacturing complex in Puebla, Mexico, and a small but troubled operation in Shanghai. If it could master lean production in manufacturing operations and product design, and transfer these techniques to its Brazilian, Mexican, and Chinese operations, its prospects might improve rapidly and dramatically. This is particularly so given its geopolitical advantage in establishing operations in Eastern Europe.

Clearly, the first step for the European companies is to master lean production in all areas of manufacture, so they can defend their home region. Otherwise, the Japanese and, surprisingly, the Americans may be the only lean producers in post-1992 Europe. (Ford has dramatically improved its manufacturing operations in Europe by transferring what it has learned from Mazda.) Only when the Europeans master lean-production methods will they be in a position to revitalize their manufacturing presence in North America and East Asia. By then, though, it may be too late.

## THE JAPANESE AND GLOBAL PRESENCE

The Japanese begin from a better position but face extraordinary global challenges as well. A brief look at Honda's strategy sheds considerable light on the problems ahead.

As is so often the case, the company that takes the boldest leap in the world market is the one in the weakest position at home. While Americans have practically made Hondas a cult product, the company was perceived in Japan as a scrappy, but somewhat eccentric, minor player. Unlike Toyota, Nissan, Mitsubishi, and Mazda, Honda had no close links to a *keiretsu* and no major business activity other than cars and motorcycles. Completely atypical for Japanese companies, it also had little interest in the truck market, limiting its offerings to a single mini-mini van.

Given its overwhelming reliance on exports, which account for about 70 percent of its Japanese production, Honda decided by the mid-1970s that it would be necessary to produce abroad. Its vulnerability to currency shifts and trade barriers was simply too great if it didn't spread its manufacturing base. Its U.S. auto assembly plant opened in 1982 but was initially only an assembly operation producing cars with perhaps 25 percent American manufacturing content versus 75 percent Japanese.

At the same time, Honda was searching for a manufacturing base in Europe. This was much harder to develop, because Honda started selling in Europe considerably after Toyota and Nissan, and even after Mitsubishi and Mazda. So it was last in the queue for the quotas established on Japanese imports into Britain, France and Italy by the early 1980s and had a very weak distribution network in the more open markets, such as Germany. With European sales of only 140,000 units in 1989, spread between five models, Honda was in a weak position to move immediately to a full-size assembly operation.

Instead, Honda pursued a stormy alliance with Rover Group, initially a state-owned company but now part of private-sector British Aerospace. After several licensing agreements in which Rover built Honda designs in England, this led to collaboration on the design of the model that became the Honda/Acura Legend and the Rover Sterling. Honda planned to sell Legends produced at Rover's U.K. Cowley plant to bolster its sales volume in Europe. However, it reportedly found the cars of unacceptable quality, even after rework at a new Honda-owned plant at Swindon in western England. So it quietly discontinued this effort shortly after it began.

The next step was to design and produce jointly a new midsize car, the Honda Concerto/Rover 200. In 1989 Honda took a 20-percent equity stake in Rover and provided a large amount of manufacturing assistance on the new Concerto/Rover assembled at Rover's Longbridge plant near Birmingham. The product was launched in Europe in late 1989 and will be followed in 1992 by a new joint product, the Syncro, produced at Honda's own assembly plant at Swindon, which Honda will open at that time.

Honda has therefore moved painfully toward a European-based manufacturing system, whose final form is yet to emerge, through a complex collaboration with Rover. Meanwhile, in the United States and Canada, it has steadily expanded its assembly plants at Marysville and East Liberty, Ohio, and Alliston, Ontario.

Capacity should reach 600,000 units by the end of 1990. Combined with its imports, Honda will probably pass Chrysler at this point to become number three in North American passenger car sales.[5]

Of greater interest for our purpose, Honda has steadily increased the North American content of its cars by adding a 500,000-unit engine plant at Anna, Ohio, and a host of wholly owned components operations nearby. It also obtains a wide range of components from traditional Honda suppliers from Japan who have opened American transplants nearby and from established American-owned suppliers. While local-content calculations are notoriously unreliable, Honda's claim that it will achieve 75-percent North American manufacturing value in its United States and Canadian assembled cars by 1992 is probably not far off the mark. ("Local content" is simply the portion of the car made in the United States. For example, the engine made in Anna, Ohio, is local, while the engine computer made in Japan is imported content.)

How companies can add engineering value is a much more interesting question when it comes to achieving global presence. Honda has taken the lead among the Japanese transplants in establishing an American engineering operation, both for products and manufacturing processes. What's more, Honda is already offering one model, the Accord coupe, which is styled and tooled in America; has a second, the Accord station wagon, in preparation; and talks of designing and engineering products from the ground up in North America by the end of the 1990s.

However, we shouldn't underestimate the scale of this task. Honda will have 700 engineers in Ohio and Michigan by 1991, seemingly a large number until one remembers that Ford and GM employ tens of thousands of engineers in Detroit. Even taking into account our findings from Chapter 5 that Honda and other Japanese firms are likely to use engineers up to twice as efficiently as the Americans, Honda has a long way to go to implement top-to-bottom lean production in North America. This process took Ford fifty years in Europe. Honda is renowned for its ability to do things in a hurry, but we shouldn't underestimate the problems involved in developing a complete product-development system on a new continent.

Even if Honda can move very rapidly, we have to ask how the company's growing global operation will be managed. Honda's public answer is that it will build an alliance of self-reliant regional companies in Japan, North America, and Western Europe, and even in Latin America (Brazil) and Southeast Asia (Thailand). The major regional companies are to conduct top-to-bottom design, engineering, and manufacture of products. These are to be sold primarily in their region of manufacture, but limited volumes are also to be exported to other regions to serve niche markets in a pattern similar to that of the hypothetical "post-national" company in Figure 8.4. The Accord coupe, now being exported from the United States to Japan and soon to be exported to Europe as well, is the first example of this process.

But how are the regions to coordinate their activities? What will a world Honda personnel system look like? Will the senior jobs at headquarters still be reserved for Japanese nationals joining the company at age twenty-two? How long will it take to bring about the alliance of self-reliant regions? These are all questions Honda must answer if it is to succeed at becoming a truly global enterprise.

## SPECIFYING THE MULTIREGIONAL ENTERPRISE

Ford and Honda, the two most advanced companies in building a multiregional production system, have made considerable progress, although neither would claim it has yet found the perfect solution. By comparison, however, the Europeans and Chrysler in the United States haven't left the starting gate, and the rest of the Japanese, including Toyota, substantially trail Honda. Clearly, the world and the auto industry have a long way to go before multiregional production is fully implemented. We'll consider this challenge from a political perspective in our final chapter. Here, we look at the management challenge these companies face by laying out the features of a truly global enterprise that can achieve multiregional lean production in the 1990s.

Our goal is to specify the ideal enterprise in much the way buyers of such craft-built cars as the Aston Martin used to specify the car of their dreams. Unfortunately, no such dream machine currently exists, so we will create it: Multiregional Motors (MRM).

The management challenge, we believe, is simple in concept: to devise a form of enterprise that functions smoothly on a multiregional basis and gains the advantage of close contact with local markets and the presence as an insider in each of the major regions. At the same time, it must benefit from access to systems for global production, supply, product development, technology acquisition, finance, and distribution.

The central problem is people—how to reward and motivate thousands of individuals from many countries and cultures so that they work in harmony. Unfortunately, the three models developed so far for this enterprise are inadequate. The first is extreme centralization of decision-making at headquarters, a headquarters almost invariably located in the home country and staffed by nationals of the home country.

As we've seen, this was Ford's approach from 1908 into the 1960s and is the approach of many Japanese companies moving offshore now. Centralization produces bad decision-making. Much worse from a political standpoint, it generates intense resentment in other regions, as it soon becomes apparent that the most important decisions are always reserved for headquarters and for those employees with the right passport.

The commonly pursued alternative has been extreme decentralization into regional subsidiaries, each developing its own products, manufacturing systems, and career ladders in isolation from the other regions. This was the position of Ford of Europe in the 1970s and still describes GM of Europe. This hermetic division by regions results in a narrow focus, ignores advantages of cross-regional integration, and creates gilded cages for highly paid national executives unable to rise any farther in their organization.

Strategic alliances with independent partner firms from each region, a variant on the last approach, is the third model. Examples include Mitsubishi with Chrysler and General Motors with Isuzu and Suzuki. (Indeed, Lee Iacocca has often talked of a Mitsubishi/Chrysler/European producer alliance, calling it Global Motors.)

Unfortunately, these arrangements leave the central question of coordination and overall management unanswered. Given this fact, it is hardly surprising that most strategic alliances in the motor industry (as differentiated from narrowly focused joint ventures such as NUMMI with specific, short-term objectives) have proved undynamic and unstable. The continuous bickering between Ford and Mazda, GM and Isuzu, and Chrysler and Mitsubishi suggest not that these arrangements need better management, but that they are unmanageable except in perfectly stable market conditions.

In this vacuum of choices, let us propose a new corporate

form we term the post-national. The key features of what we call Multiregional Motors are as follows:

• *An integrated, global personnel system that promotes personnel from any country in the company as if nationality did not exist.* Achieving this goal obviously will require great attention to learning languages and socialization and a willingness on the part of younger personnel to work for much of their career outside their home country. However, we already see evidence that younger managers find career paths of this type attractive.

We have met numerous Japanese managers at the U.S. transplants who look forward to long U.S. tours and to future assignments in Europe. Unlike older managers who often lack language skills, they see this path both as an interesting way to live and as the surest route to success in their company.

Similarly, Ford of Europe has recently had considerable success in recruiting European managers who do not expect or want to work in their home country and who anticipate serving for considerable periods in the United States as well. And we are now running across a number of Americans eager to work in Japan.

• *A set of mechanisms for continuous, horizontal information flow among manufacturing, supply systems, product development, technology acquisition, and distribution.* The best way to put these mechanisms in place is to develop strong *shusa*-led teams for product development, which bring these skills together with a clear objective.

In most Western companies, much activity is unfocused. Product planners work on products that never get the green light, massive amounts of staff waste time fighting fires. The best Japanese companies, by contrast, believe strongly that if you aren't working directly on a product actually heading for the market, you aren't adding value. So involving as many employees as possible in development work and production is vital. Companies should keep their eyes on the product the consumer will buy.

Teams would stay together for the life of the product, and team members would then be rotated to other product-development teams, quite possibly in other regions and even in different specialties (for example, product planning, supplier coordination, marketing). In this way the key mechanism of information flow would be employees themselves as they travel among technical specialties and across the regions of the company. Everyone would stay fresh and a broad network of horizontal information channels would develop across the company.

Teams in Japan do stay together now, but members aren't assigned to new projects in new regions as a way to create a global flow of horizontal knowledge and to give every employee a sophisticated understanding of the world. (The question, of course, is not whether this is a good idea in principle, but whether enough employees would find it attractive.) As they travel through the company and across regions, these managers also would create a uniform company culture—a largely implicit way of thinking and doing things that every organization needs to function well.

• *A mechanism for coordinating the development of new products in each region and facilitating their sale as niche products in other regions*—without producing lowest-common denominator products. The logical way to accomplish this goal is to authorize each region to develop a full set of products for its regional market. Other regions may order these products for cross shipment as niche products wherever demand warrants.

Since MRM will ship products in roughly equal volumes among its regional markets, it can largely ignore currency shifts: Losses on cars shipped in one direction are offset by higher profits on cars sent in the other direction.

When currencies shift today, management typically panics and explores ways to relocate production to low-cost areas quickly. Or they search for trade protection.

MRM's managers, who will have a long-term commitment to a world-class lean-production system in each major region, can be much more relaxed, provided that one additional element of the post-national enterprise is in place: internationalized financing and equity.

Most of today's motor-vehicle companies have the bulk of their equity and borrowings in their home region and pay dividends and loans in their home currency. So shifts in currency values are still a problem, even if they have succeeded in establishing a multiregional production system.

Consider an American firm with dollar-based loans and dividends. A strengthening of the dollar could prove very damaging if the company earned the bulk of its profits offshore—even though the company's market position and profitability in local currency terms in all three regional markets was unchanged.

Internationalizing corporate equity so that funds are raised in each region in rough correspondence with sales volume and manufacturing investment would largely eliminate this concern. Dividends could then be paid in the currency of the region to insulate the organization from currency shifts between regions.

With these new approaches to personnel management, information flows, product development, cross-regional trade, and internationalized finance in place, it may be possible to create an MRM appropriate to the regional world of the 1990s. We believe it is particularly important that motor-vehicle companies such as MRM come into being, not just for commercial reasons but because of the emerging global political challenge. We will return to this point in the final chapter.

# DIFFUSING LEAN PRODUCTION

We have now gone through the elements of lean production in the factory, in product development, in the supply system, in the sales and service network, and in the hypothesized multi-regional lean enterprise. Our conclusion is simple: Lean production is a superior way for humans to make things. It provides better products in wider variety at lower cost. Equally important, it provides more challenging and fulfilling work for employees at every level, from the factory to headquarters. It follows that the whole world should adopt lean production, and as quickly as possible.

As with so much else, however, this is easier to say than to do. Whenever a fully developed set of institutions is firmly in place—as is the case with mass production—and a new set of ideas arises to challenge the existing order, the transition from one way of doing things to another is likely to prove quite painful. This is particularly true if the new ideas come from abroad and threaten the existence of major institutions in many countries, in this case the massive home-owned, mass-production motor-vehicle companies. With help from their governments, these institutions may resist change for decades or even overwhelm the new way of thinking.

So we are not certain that lean production will prevail. We do believe that the 1990s will tell the tale. We *are* convinced that the chances of lean production prevailing depend critically on a wide public understanding of its benefits and on prudent actions by old-fashioned mass-producers, by the ascendent lean producers, and by governments everywhere.

In the remaining chapters, we'll shift from analysis—what lean production is and where it came from—to prescription. We'll present a vision of how the world can make the transition to a new and better way of making things with a minimum amount of pain and tension.

# CONFUSION ABOUT DIFFUSION

## 9

Between 1914 and 1924, Henry Ford's and Alfred Sloan's industrial innovations destroyed a vigorous American industry, the craft-based motor-vehicle business. During this period, the number of U.S. automobile companies fell from more than 100 to about a dozen; of which three—Ford, General Motors, and Chrysler—accounted for 90 percent of all sales.[1]

Yet there was no panic, no protest, no call for government intervention. True, a series of social critics questioned the new type of factory life that mass production was introducing, but no one called for the protection of the embattled craft producers.

The reason for the lack of resistance is not far to seek. Even as Ford and Sloan were demolishing one industry they were creating a second—the mass-production motor industry, and they were doing it in the same city where craft production had flourished most vigorously. The growth of this second industry was so dramatic that practically all the skilled workers of the craft-based industry could find jobs building tools and doing other skilled tasks to support the mass-production system. Indeed, until 1927, when Model T sales collapsed, Henry Ford faced the continual problem of finding enough skilled workers in the Detroit area to man his toolmaking operations. Meanwhile, the rapid growth of car and truck sales, combined with continually falling prices, was creating hundreds of thousands of new unskilled jobs on the assembly line.

In addition, Ford and Sloan were Americans—hometown boys, even—and Henry Ford adroitly portrayed himself as a folk hero bringing a high standard of living to the common man. There was no suggestion of foreign menace in the triumph of mass production in Detroit.

No one has repeated Ford's and Sloan's ease in supplanting one type of production method with another. Indeed, as soon as mass production began to move abroad from the United States, it began to encounter resistance. This pattern is repeating itself today as lean production displaces mass production. The basic problem was and is that existing companies and workers using older production techniques find it hard to adopt new ways pioneered in other countries. The alternative method of diffusing new techniques—the arrival of foreign companies—tends quickly to set loose a nationalistic reaction in the countries where the old-style companies are based. The result has often been a delay of decades in substituting new production methods for old ones.

## MASS PRODUCTION ENCOUNTERS CRAFT PRODUCTION IN BRITAIN

In October 1911, Henry Ford opened an auto assembly plant at Trafford Park near Manchester, England.[2] With the exception of a small assembly plant in Windsor, Ontario, just across the Detroit River from his Highland Park plant, this was Ford's first venture abroad. Ford built the Trafford Park plant to overcome the limitations of the transport of the day, but soon he needed it to overcome trade barriers as well.

In 1915, Britain had abandoned free trade and adopted the McKenna tariff, which imposed a 25-percent tax on completed autos that came from abroad. (Most of these imports came from the United States.) Parts, by contrast, bore only a 10-percent tariff, so foreign manufacturers had a strong incentive to establish final assembly plants in England.

Initially, everything went well at Trafford Park. Ford dispatched a large number of American managers from Detroit to replicate exactly the mass-production system he was perfecting at Highland Park. When workers were hired, they were explicitly told that they would be handymen—that is, none of their craft skills, if they had them, would be needed, and they would be available to move from job to job in the assembly hall.[3] (Indeed, one manager at Trafford Park estimated that it took only five to ten minutes to train a worker to do practically any of the assembly jobs in the plant.) The first powered assembly line was installed in September 1914, about nine months after the first powered line became a reality at Highland Park. By 1915, the full complement of Ford assembly technology and techniques was in place at Trafford Park.

The implications of Ford's mass-production system were not lost on the skilled workers Ford was hiring for the bodybuilding shop. The upholstery department, for example, used special molds to eliminate the skilled job of hand stuffing. Seat-cover sewing was standardized and simplified. In the body shop, stamping presses eliminated the skilled panel beaters (whose descendants we recently encountered at Aston Martin). A paint-spraying system substituted for the skills of the brush painter. The result in 1913 was a strike that closed the body shop, as skilled coach builders protested Ford's methods and argued for a return to skilled work paid by the traditional piece-rate system.[4] (Ford had broken ranks with British employers and paid by the hour at Trafford Park, as he did in Detroit.)

Because workers doing simplified tasks on the production line could easily be replaced, and Ford, in any case, could supply finished car bodies from Detroit, the strike soon collapsed. It was more expensive to ship car bodies from Detroit—the British tariff and damage in transit drove costs up—but Ford could do so until the strikers exhausted their savings and gave in. By 1915, no one was challenging Ford's system in the plant, and a Ford manager from Detroit reported that productivity at Trafford Park was comparable to that at Highland Park.[5] Seemingly, mass production had triumphed in a new setting. Logically, it should soon have become the dominant form of production in England and perhaps in Europe as well.

However, this was not to prove the case. The reason was a series of events that make us very cautious about the rapid and easy triumph of lean production in the 1990s.

## THE TRIBULATIONS OF MASS PRODUCTION IN BRITAIN

Ford's problems began in 1915 with an improbable event, the mission of his Peace Ship.[6] Ford was an ardent isolationist: The United States should stay out of the First World War, he protested. To that end, he organized a trip to Europe aboard a chartered ship to broker peace privately between Germany and Britain. In Britain, however, the public perception of Ford's motives was emphatically different from Ford's own statements: He was universally thought to be pro-German. The result was popular resistance to Ford products—many newspapers, for example, refused to accept Ford's advertising—and a loss of morale among Ford's British employees.

Energetic action by Ford's British managers offset some of the bad feeling, but Ford's problems soon multiplied. The energy and horsepower taxes enacted after the war were particularly hard on Ford products. The horsepower tax in particular, which was proposed to the government by Ford's competitors and which favored their "long stroke" designs over Ford's short-stroke engine, proved a crippling blow. Ford's Model T, envisioned as the "universal" car, was soon the wrong car for Britain. The consequence of Ford's bad fortune was that his factory often ran at a fraction of its capacity and the company back in Detroit seemed to lose interest in its performance.

Not surprisingly, factory performance seemed to deteriorate steadily. None of the English managers shared a conception of management that was compatible with mass production. The idea of a manufacturing career beginning on the shop floor with hands-on management was unattractive to middle-class Englishmen who emerged from an educational system that steered them toward the civil service, the law, and other types of high-level administration. They didn't want to get involved with the nitty-gritty of running anything. Rather, they wanted to delegate operational details, just as they did with the Empire.

In addition, British managers were persuaded that Englishmen, with a long experience of craft working, would not tolerate Ford's methods. For a short period perhaps—under the whip hand of the American managers—but certainly not over the long term.

Consequently, managing the shop floor soon became, by default, the responsibility of the shop steward, who typically was a skilled craftsman highly suspicious of mass production. These front-line managers lobbied to retain traditional skills and piece-rate payment systems, which made no sense in continuous-flow production, where every worker's effort is paced by that of every other worker.

The performance of Ford's English plants went backward to a point where an enormous gulf developed between practice in Detroit and in Trafford Park. When Ford abandoned Trafford Park and established a full-fledged, top-to-bottom manufacturing system at Dagenham, England, in 1931, the gap became even greater. Moreover, it persists to this day.

With the Ford Motor Company—the inventor of the new system and the industry leader—making such a poor showing, it's hardly surprising that Ford's English competitors embraced mass production with only partial success.

## INDUSTRIAL PILGRIMS: THE TRIP TO HIGHLAND PARK

By the spring of 1914 Henry Ford was actually turning out two products at Highland Park: Model T's and refurbished captains of industry. An endless stream of industrial pilgrims began to arrive around 1911 in a flow that continued for forty years. (Indeed, the pilgrimages ended only with the visit of Eiji Toyoda in 1950.) The Ford Archive in Dearborn, Michigan, contains an extraordinary gallery of pilgrims photographed with the master.

They range from André Citroën (Citroën), Louis Renault (Renault), and Giovanni Agnelli (Fiat), to nameless Russians anxious to add mass-production techniques to Lenin's formula of "Soviets plus electrification equals Communism." (Lenin later amended this formula to "Soviets plus Prussian railway administration plus American industrial organization equals Socialism.")[7] A particularly striking photo taken in 1921 captures Charlie Chaplin and Henry Ford in smiling mutual admiration alongside the Highland Park assembly line, at a time when Ford was still perceived as a miracle worker for the masses rather than the enemy of labor.[8]

William Morris, founder of the Oxford Motor Company (and its subsidiary MG), and Herbert Austin, founder of the Austin Motor Company, were among these pilgrims. After Morris visited Highland Park in 1914, he returned to England determined immediately to copy mass-production techniques in his own factory. But he didn't have an easy time of it.

The war interrupted production, and a hand-powered assembly line was not in place until 1919. In this arrangement, cars rode on dollies on rails. However, the automobiles were pushed along to the next station by hand, so the whole line ran at the speed of the slowest worker. The assembly line was not powered until 1934, twenty years after Ford's first powered line in Detroit. Morris also found it difficult to divide labor to the extent Ford had. For example, his final assembly line consisted of eighteen separate work tasks in 1919, while Ford's had forty-five by 1914. Finally, Morris found it difficult to develop managers who were willing and able to operate a Ford-style mass-production system.

Quite incredibly, Morris continued to pay all his workers on a piece-rate system until after World War II, despite the fact that all tasks were linked together in a continuous line. The workers' only concern, naturally, was to work as fast as they could to meet the day's quota and qualify for a bonus, then knock off. We can easily imagine the consequences of this system for the quality of the finished product.

Morris stuck with piece rates, because he could think of no other way to get his employees to work. His weakness in line management meant that he could run his plant only indirectly with the aid of shop stewards who mediated between Morris and his workers on the work pace and piece rates. In short, while he was trying to create an exact copy of Highland Park, he was achieving instead an imitation of Ford at Trafford Park after the American managers left.

In frustration, Morris gambled on what we would today call high tech. In his engine plant, he proposed to totally automate the machining of engines, flywheels, and transmissions in order to eliminate most workers, both skilled and unskilled. Accounts of his experience, as we'll see, read very much like the experiments of General Motors and Fiat with advanced automation in the 1980s, automation which was adopted out of similar frustrations.

When Morris installed his equipment in 1925, he found that he could capture considerable savings as workers moved engine blocks and transmission cases along rails from machine to machine. Then each machine would work more or less automatically to complete a given task. (Previously, the machines had been grouped by type—all milling machines in one area, grinding machines in another area, and lathes in a third area—and the parts had been carried one at a time on carts, a procedure that involved lifting and tugging at every machine.) However, what the technology *couldn't* do was totally eliminate human intervention by automatically transferring the parts from one machine to the next. Indeed, this goal still exceeds the capabilities of technology.

Herbert Austin had a very similar experience, except he never contemplated a high-tech leap as a way out. After visiting Highland Park in 1922, he returned to England determined to copy Ford's system. He succeeded halfway. Austin installed assembly lines, although these were not powered until 1928 or later, and divided work into small unskilled tasks. But shop-floor management was still very weak, so he stuck with piece rates as the best way of motivating his workers.

An Austin worker years later told a BBC interviewer how this system worked in practice: "Well . . . you had so much time for doing a job. If you were on ordinary time you'd get two pounds a week and then you'd have to go quicker and quicker to get more. So the [assembly line] would start at time and a quarter . . . and then it would go up to time and a half, or three pounds a week. [Management] would speed it up as we got used to it . . . to double time. And when it got to double time, they'd stop. No more, no faster. And what we used to do, we'd have a real good go, we'd pick the bodies up . . . and jump the pegs [to move the cars along faster than the assembly line] and we'd make it up to double time and a half, which was about five pounds a week, and this is a lot of money in them days."[9]

The idea of workers running down the line carrying automobiles faster than the conveyor seems comical today. And the system must have had horrendous consequences for the quality of the finished car, but Austin could see no other way to manage. As one of his senior managers argued, in defense of the piecework pay system: "Some form of extra wage [the bonus] must be paid to a man if he is expected to work harder. The only alternative is to pay a high wage similar to the Ford system and insist on task achievement. . . . The daily task system at fixed wages may perhaps, be workable in American . . . factories, but the necessary . . . driving works policy would not be acceptable either to English Labour or Management." (The English term for factory is "works," so a "driving works policy" is simply machine pacing with standard day rates.)[10]

The consequence of this hybrid system, now called the British system of mass production,[11] was that British plants, including those of General Motors and Ford, never matched the productivity or quality of U.S. plants. Indeed, it was not until the financial crisis of 1980, sixty-seven years after the introduction of the powered assembly line at Highland Park, that Rover (formerly British Leyland), the successor to the merged Austin and Morris companies, finally adopted standard hourly rates and explicitly set out to match the productivity of the Americans. (British Leyland was nationalized in 1975. By 1979, it was hemorrhaging red ink, and a new management was installed with clear instructions to make the company efficient or close it down.) By this time, of course, American-style mass production was already under siege from Japanese-inspired lean production.

## MASS PRODUCTION IN CONTINENTAL EUROPE

The French, German, and Italian experience with mass production was a variation on the English theme, with the difference that the Americans had a harder time spreading their homegrown system through direct investment. Citroën, Renault, and Agnelli, to cite the three industrialists most taken with the mass-production concept, struggled through the 1920s and 1930s to implement the idea in chaotic economic and political conditions. Their problem stemmed partly from resistance by craftspeople, but also from the lack of a stable domestic market as European economies careened from hyperinflation to depression.

Ford tried to lead through example with investments at Cologne in West Germany and Poissy near Paris, and GM bought the tiny German producer Opel in 1925. However, Italy firmly closed its door to both companies. What was more, the need of Ford and GM to produce practically every part for every car within each European country, due to trade barriers within Europe and across the Atlantic, pushed up costs, restricted the size of the market, and, in general, retarded the spread of mass production. As Europe plunged into war once more at the end of the 1930s, the progress of mass production had been quite limited. In turn, the failure of the European economy to grow was one of the underlying causes of the war. That is, because mass production hadn't progressed, the European economy stagnated, creating the conditions that helped lead to war.

After the war, change became very rapid. Much of the European economic miracle of the 1950s and 1960s was nothing more than a belated embrace of mass production. Volkswagen built Wolfsburg as the world's largest car plant under a single roof, and Renault and Fiat responded with Flins and Mirafiori, all plants included in our survey reported in Chapter 4.

By the mid-1960s, continental Europe had finally mastered American techniques (just as Eiji Toyoda and Taiichi Ohno were moving beyond them) and began to challenge Detroit in export markets.[12] At the same time, the Americans were investing aggressively in Europe and had developed complete production development and supplier systems on a Europe-wide basis. The process of replacing craft production with mass production had been completed, but it had taken fifty years.

## LEAN PRODUCTION ENCOUNTERS MASS PRODUCTION

We have paid careful attention to the substitution of mass production for craft production because of the perspective it gives on the current challenge of superimposing lean production on mass production. In fact, the new challenge seems much greater.

In Europe in the 1920s, the craft-based auto industry was quite small. The substitution of mass production, had it succeeded, would certainly have increased employment spectacularly, as it did when mass production finally arrived in mature form in the 1950s. Nevertheless, the threat of foreign domination (by the Americans) was so frightening and the mismatch between existing institutions and conceptions (such as English notions of management and continental notions of skilled work) so great that Europe sealed itself off rather than adapt.

In the 1990s, the fear of foreign domination (this time by the Japanese) will surely prove as great. However, the mature nature of the motor-vehicle market in North America and Europe, coupled with the efficiency gains inherent in lean production, mean that no painless solution can emerge. As lean production replaces mass production but the same number of cars and trucks are built each year, many jobs will disappear.

What's more, the current Western automotive work force is in precisely the opposite position of craft workers in 1913. The introduction of mass production created new jobs for craft workers—these workers made the production tools needed by the new system. By contrast, lean production displaces armies of mass-production workers who by the nature of this system have no skills and no place to go.

## THE THREAT ON THE HORIZON: INITIAL MISPERCEPTIONS OF LEAN PRODUCTION

Anything that is new is likely to be misunderstood, typically by attempts to explain the new phenomenon in terms of traditional categories and causes. So as the industrial revolution Toyoda and Ohno had wrought began to make itself felt abroad, through exports of finished units, what they had achieved was widely misinterpreted.

One popular explanation in the 1970s of the Japanese success was simply that Japanese wages were lower, an explanation that fits nicely into established theories of international trade. A second explanation was summed up by the phrase "Japan, Inc." This theory attributed Japanese success to its government's protection of the domestic market and its financial support for Japanese car companies, through tax breaks and low interest rates, as they tried to target export markets. A third explanation was high tech, notably the widespread adoption of robots in the factory. Together, these made the emergence of Japan understandable but also sinister—artificially low wages combined with government financial support (for example the 1970s tax laws promoting the installation of robotics)—to beat the Western mass-producers at their own game.

What's more, there were elements of truth in each of these explanations. Japanese wages were substantially lower than American wages prior to the currency shifts of the 1970s. The Japanese government's efforts to protect the domestic market and domestic ownership were absolutely essential to the initial growth of the Japanese industry. And the level of automation on average in Japan by the early 1980s was higher than in the West. What these explanations could not explain was how the Japanese companies continued to advance in the 1980s despite currency shifts and a massive movement of operations offshore, where MITI was of little help. Nor did they explain why the Japanese firms gained major benefits from automation while Western firms often seemed to spend more than they saved. Deeper explanations of these mysteries required an understanding of lean production.

# CONFUSION ABOUT DIFFUSION

## THE NEW INDUSTRIAL PILGRIMS: THE TRIP TO HIROSHIMA AND TOYOTA CITY

Fortunately, a new pilgrimage route soon emerged, this time from Detroit to Japan. Most notable of the early pilgrims was a joint group from the Ford Motor Company and the United Automobile Workers Union, reversing the steps Eiji Toyoda took in 1950.

In 1980, the Ford Motor Company suffered what turned out to be a very timely crisis. The company began to lose both vast amounts of money and large chunks of market share. Fortunately, senior Ford management and the UAW leadership at Ford realized that the problem was not primarily cyclical, although the 1980 downturn in the market was the worst since the 1930s. They concluded that the Japanese competitors were doing something fundamentally new—in short, that the traditional explanations we just cited were inadequate to account for the Japanese success.

They decided to go to Japan to see for themselves, a journey that became feasible by Ford's purchasing 24 percent of Mazda in 1979. This meant that senior Ford executives and Ford's UAW leadership could gain full access to Mazda's main production complex in Hiroshima and determine for themselves why Ford was taking a beating in international competition.

Ford had a second stroke of luck in its Mazda connection, notably that Mazda itself had experienced a timely crisis in 1974. The failure of its technology-driven product strategy—based on its use of the fuel-hungry Wankel engine—caused Mazda to transform its production complex at Hiroshima into a faithful copy of Toyota's lean-production system at Toyota City. If the Ford and UAW executives had visited Hiroshima in 1973 rather than 1981, they might well have reached the wrong conclusions.

After several weeks in Hiroshima, followed by months of careful staff work, the Ford executives and Ford's UAW leaders discovered the answer to Japanese success: lean production. Specifically, they found that Mazda could build its 323 model with only 60 percent of the effort Ford needed to manufacture its Escort selling in the same market segment. Moreover, Mazda made many fewer manufacturing errors in doing so. Equally striking, Mazda could develop new products much more rapidly and with less effort than Ford and worked much more smoothly with its suppliers to do so.[13]

Back in the United States, the severe crisis at Ford—which by

1982 was threatening the survival of the company—was breaking the logjam of old thinking and entrenched interests. Suddenly, employees at all levels of the company were ready to stop thinking about how to advance their careers or the interests of their department and to start thinking about how to save the company. This situation is the very definition of a creative crisis, and the news from the pilgrims to Hiroshima came at just the right time. During the 1980s Ford was able to implement many elements of lean production and the results soon showed up in the marketplace.

Chrysler, meanwhile, was in much deeper trouble than Ford or GM and was already a ward of the U.S. government. Why it failed to learn much about its real problems in a time of crisis, despite its equity tie-in and access to Mitsubishi, is a tragic mystery.

General Motors' experience was quite different from Ford's. The company was also represented along the pilgrimage route, but until recent years it lacked the crisis needed in any mass-production company to take the lessons of lean production to heart. GM was rich in 1980. Although it lost $1 billion in 1982, it still had little debt and was, by far, the world's largest company. It dealt with its problems mainly by retreating from market segment after market segment and attempting dramatic leaps in productivity through the introduction of all available new production technology when it introduced new models such as GM-10. No one complained as the Japanese moved in to fill the competitive gap until quite recently, when institutional investors started becoming nervous that GM was slowly liquidating itself.

In the 1980s, GM's primary means of education was through the planning process for its joint venture with Toyota in California. As this plan became a real possibility in 1983, senior GM executives spent a lot of time at Toyota City. As Jack Smith, now GM's vice-chairman, noted later, "It was the first time we really had a clear understanding of how they ran. . . . The data [on productivity] was just unbelievable."[14]

As we showed in Chapter 4, the NUMMI joint venture was an extraordinary success. However, transferring the lessons learned throughout the vast General Motors organization has proved hard work. The fundamental problem is that making the transition from mass production to lean production changes the job of every worker and every manager. What's more, in the absence of market growth, many jobs are eliminated. Since GM didn't face a crisis

## CONFUSION ABOUT DIFFUSION

in the 1980s and failed to find any opportunities for growth, it was simply not able to face up to the challenge.

For the same reason, the European motor-vehicle companies have been represented only modestly along the pilgrimage route to Toyota City and lean production. The European auto market was vigorous in the last half of the 1980s, setting a new sales record each year from 1985 through 1989, and Japanese competition was contained through formal trade barriers and a welter of gentlemen's agreements.[15] As a result, the European companies had little external pressure on them to change. As mentioned, the most notable movement toward lean production in Europe was not by European companies but by the American company, Ford, which attempted to apply in Europe what it had learned in Japan.

An experience of our group perfectly summarizes this situation. In 1982, while visiting a French auto assembly plant in the Paris area, we encountered a young engineer. He had just returned to the plant after a year-and-a-half exchange visit in a Japanese car company in Japan. He was bubbling over with enthusiasm about the contrast between lean production, as he had discovered it almost by accident in Japan, and the mass-production practices of his own company. He was eager to introduce lean-production techniques as quickly as possible. His main concern was where to begin and how to capture the attention of senior management.

Our discussion was cut short by a remarkable event—a violent industrial action involving two factions of the North African guest workers who held practically all the production jobs in the plant. These workers were represented by two separate unions and were embroiled in a dispute over work rules. As the tension between the two factions grew toward a confrontation in which a large number of finished vehicles were vandalized, the plant managers advised our team to leave. We wished the young engineer the best of luck in implementing lean production as we hurriedly departed.

In the fall of 1989, quite by accident, we encountered the same engineer at one of the provincial plants of his company where he was now head of manufacturing. We asked what had become of his efforts to institute lean production. For a moment he looked puzzled, but then he remembered our initial encounter and gave us a remarkable reinterpretation of events: The real problem, he had concluded, was the guest workers in the French auto plants in the Paris area. In the provinces, however, guest workers were not an issue. All the workers were French, a spirit of cooperation prevailed, and he would stack up his current plant against any in the world.

We had considerable difficulty in continuing the conversation at this point, because the survey we had just completed showed that his plant takes three times the effort and makes three times as many errors as the best lean-production plants in Japan in making a comparable product. What's more, the amount of space and the inventory levels in his plant were several times the Japanese level, and the French plant was focused on a single product in one body style on each of its production lines.

In short, since his company hadn't faced a challenge from a lean producer, he wasn't able to initiate the change in mind-set needed to implement lean production. The young bearer of the message had returned from the pilgrimage and fitted himself into the familiar industrial landscape of mass production. We felt a profound sense of gloom as we left the plant.

## BENCHMARKING THE PATH TO LEAN PRODUCTION

We in the IMVP have been pilgrims ourselves, first to the best lean-production facilities—all in Japan until very recently—and then back to the strongholds of mass production in North America and Europe. We believe we have traveled farther and made more comparisons than anyone else, either inside or outside the motor-vehicle industry. So, where do we stand along the path to global diffusion in lean production? And what must happen for the whole world to embrace this system?

Remember that as a practical matter there are only two ways for lean production to diffuse across the world. The Japanese lean producers can spread it by building plants and taking over companies abroad, or the American and European mass-producers can adopt it on their own. Which of these method proves dominant will have profound implications for the world economy in this decade.

## DIFFUSION THROUGH JAPANESE INVESTMENT IN NORTH AMERICA

Japan's move offshore began as a trickle in the 1960s. Its first major initiative was Nissan's engine and assembly plant in Mexico in 1966. Not much else happened for a long time—unless you count extremely low-volume assembly plants ("kit plants," in car talk), generally run by licensees rather than the Japanese company itself, in protected developing-country markets. For example, in 1966 when the Brazilian government prohibited further imports of complete vehicles, Toyota licensed a local Brazilian company to assemble kits of parts for its Land Cruiser utility vehicle.

Honda made Japan's first serious foreign investment with its Marysville, Ohio, complex, which began assembly in 1982. Once one company was firmly committed offshore—and as it became apparent that shifting currencies and persistent trade barriers (for example, the Voluntary Restraint Agreement on Japanese finished cars entering the United States) made foreign investment inevitable—all the Japanese companies rushed to follow Honda's lead to North America.

The large number of Japanese auto companies (eleven) and the intensity of their rivalry has led to an extraordinary investment boom, as shown in Figure 8.2 in the last chapter.

The assembly plants came first, followed by engine plants, and now by a wide variety of parts plants. What's more, the investment flow is still broadening. Honda, Nissan, and Toyota have announced plans to design and engineer complete vehicles in North America by the late 1990s. With that step, they'll complete the process of constructing a top-to-bottom manufacturing system. The other Japanese companies are sure to follow.

The speed and scale of this process are truly extraordinary. Indeed, nothing like it has ever occurred in industrial history. In effect, between 1982 and 1992 the Japanese will have built in the U.S. Midwest an auto industry larger than that of Britain or Italy or Spain and almost the size of the French industry. By the late 1990s, the Japanese companies will account for at least a third of North American automobile production capacity—perhaps much more—and have the ability to design and manufacture entire vehicles in a wholly foreign culture 7,000 miles from their origins.

What's more, politics permitting, these investments will continue until the American companies revitalize their operations and stand their ground in the marketplace or are eliminated.

By contrast, Ford established an initial assembly plant in Europe in 1911, added full manufacturing at two sites—Dagenham, England, and Cologne, Germany—in 1931, and completed the process with a full product-development team in 1961. It took fifty years for Ford to accomplish what the Japanese may achieve in fifteen years. General Motors was no quicker. It bought the tiny Opel company in Germany in 1925 but only moved to high-scale production after World War II and did not implement a full product-development system until the mid-1960s. Chrysler didn't attempt foreign assembly and manufacture at all until the late 1960s and soon vanished from the scene. The company's crisis in the late 1970s forced it to sell off its European operations.[16] Even so, Europe was abuzz in the late 1960s with talk of the "American challenge" in which the American multinationals were seen as threatening to take over the entire European motor industry.[17]

## JAPANESE DOES NOT EQUAL LEAN

In the excitement over the transplants, many people seem to forget a point we stressed in Chapter 4: that all transplants in North America do not perform at the same level. The best performing plant, Company Y, took 18.8 hours to perform our standard assembly tasks on our standard car and needed about 5 square feet of factory space per year per car. A competing transplant located nearby, Company Z, needed 23.4 hours per car and used more than 13 square feet of factory space, by far the least efficient use of space in our entire world sample.

Both of these plants are Japanese, but one is much leaner than the other. What explains this very substantial difference in performance?

One reason: Company Z is not so proficient at lean production in Japan. Its plants there also trail the performance of Company Y's. Again, we must stress that "lean" does not equal "Japanese." While average Japanese performance is very impressive, a few Japanese companies seem to have been inspired more by Henry Ford than Taiichi Ohno, while a few companies in the West—ironically, the Ford Motor Company is the best example—have greatly modified their factories and come close to leanness in the 1980s.

A second reason for the difference in performance between the best and worst transplants is that Company Z delegated most of the operational aspects of its plant, including design and layout, to Americans lured away from Detroit. This approach is likely to carry grave risks of the sort we mentioned when we looked at Ford's attempt to transfer mass production to Europe in 1911—namely, managers, not fully understanding and committed to the company's production system, may not be able to introduce or sustain lean production in a new milieu. High-performing company Y's approach has been to send a large number of experienced managers from Japan to run its U.S. plant, and it is getting superlative results, exactly comparable, in fact, to the company's performance in Japan.

The difference, we must emphasize, is not that Company Y's managers are Japanese—or any other nationality, for that matter—but that they collectively possess many years of experience and know-how in making lean production work consistently in assembly plants. As a senior executive from Company Y emphasized in an interview, "We believe that our production system, with its many nuances, can be learned by anyone . . . but it takes ten years of practice under expert guidance."

If one accepts this manager's estimate of the time and personnel it takes to transfer lean production—and we do—it follows that the best Japanese companies may be constrained in how quickly they can build up foreign production operations. With only so many experienced managers with the language skills needed to operate in foreign environments, it may never be possible for these companies to go as fast as they would like in opening new plants.

Equally significant, foreign governments may slow the efforts of the Japanese companies by placing restraints on the number of expatriate managers they allow to work in their countries. The U.S. government, for example, has taken an increasingly stringent line on Japanese nationals managing the transplants, apparently from the conviction that the purpose of transplants is to create jobs for Americans. So it is naive to assume that lean production can be transplanted instantly by the Japanese—just as it is naive to think that all Japanese companies will be equally lean and competitive as they move abroad. Indeed, because of weaknesses in product design and marketing, it is even possible that some of the transplants recently opened by the weaker Japanese firms may fail.

## DIFFUSION THROUGH LEARNING BY THE AMERICAN FIRMS

But what of the Americans? Where do they stand along the transitional path to lean production? Without question, the U.S. industry as a whole is getting better at factory operations. Every company has improved considerably. However, GM and Chrysler have improved their operations largely by simply closing the worst plants, like GM Framingham, rather than by improving every plant. The Chrysler St. Louis 1 assembly plant illustrates this process.

St. Louis 1 has been assembling 210,000 Dodge Daytona and Chrysler LeBaron models with a work force of 3,400. The best Japanese transplants can assemble the same number of cars with about 2,100 workers. Chrysler and its union faced a simple choice: Convert from mass to lean production while displacing 1,300 workers, or close down altogether. Neither the company nor the union found a way to make the transition to lean production, and the plant will close at the beginning of the 1991 model year.

This outcome has recurred repeatedly at GM and Chrysler over the past three years, as shown in Figure 9.1. The two companies have together shut down nine North American plants, while fully converting none to lean production.[18] As this process continues, GM and Chrysler are enveloped in an ever-deepening sense of gloom in which a slow retreat never quite seems to trigger the crisis that might shake free the logjam of outdated management thinking and union-company relationships and lead to revitalization.

Ford, as we've seen, has made good use of its 1981 crisis and the pilgrimage to Hiroshima and found ways to equal the productivity of the transplants. We are concerned, however, that its plant-level performance is best when it can focus its factories on a single model with only a few options. In plants with a complex model mix, Ford's performance is much less impressive. So the company has traveled only part of the way down the path to leanness even in the factory. Nevertheless, Ford has made a bold start and has bought time to perfect its own version of lean production.

As we saw in Chapter 6, the Americans have begun to rationalize their supplier system. The number of suppliers to each company has been reduced dramatically and the attitude toward

# CONFUSION ABOUT DIFFUSION

**FIGURE 9.1**

**North American Assembly Plants Opened or Closed by the American-Owned Automobile Companies, 1987–1990**

| Company | Plant | Year Closed | Capacity |
|---|---|---|---|
| *Closings* (10): | | | |
| GM | Detroit, Michigan | 1987 | 212,000 |
| GM | Norwood, Ohio | 1987 | 250,000 |
| GM | Leeds, Missouri | 1988 | 250,000 |
| Chrysler | Kenosha, Wisconsin | 1988 | 300,000 |
| GM | Pontiac, Michigan | 1988 | 100,000 |
| GM | Framingham, Massachusetts | 1989 | 200,000 |
| GM | Lakewood, Georgia | 1990 | 200,000 |
| Chrysler | Detroit, Michigan | 1990 | 230,000 |
| Chrysler | St. Louis, Missouri | 1990 | 210,000 |
| GM | Pontiac, Michigan | 1990 | 54,000 |
| *Openings* (1): | | | |
| GM (Saturn) | Spring Hill, Tennessee | 1990 | 250,000 |
| **NET CAPACITY REDUCTION, 1987–1990** | | | **1,756,000** |

quality fundamentally transformed. Nevertheless, much remains to be done.

We also found clear signs of the intent to move toward leanness in the area of product development. Unfortunately, no product launched to date has benefitted from a truly lean development process, and one must wait until a product is fully developed and introduced in the market before drawing conclusions about its development. Program improvements instituted in 1990 will be clearly apparent only if the new model appears in 1993 or 1994— that is, in three to three and a half years rather than in the typical five—and with a greatly reduced level of engineering office. However, we have seen no clear evidence so far that the American companies can close the gap with the best Japanese companies in terms of development time and effort, only that they can dramatically shrink traditional effort levels and time frames even as the best Japanese companies continue to shrink theirs.

Five years ago, the Japanese auto makers considered forty-two months a satisfactory development pace. Today, the best companies are talking about twenty-four months as a reasonable target. So leanness continues to progress and the trailing Western mass-producers will need to move quickly indeed to catch up.

## THE BOTTOM LINE FOR NORTH AMERICA: A DECADE OF INTENSE TENSION

From one perspective, the transition to lean production is proceeding with remarkable speed and smoothness—the transplants have shown that lean production can thrive in North America, and some of the American companies show signs of mastering the new system as well. What's more, despite widespread predictions of an overcapacity crisis, the advance of the transplants has thus far been almost perfectly synchronized with the retreat of the American-owned firms. Between 1987 and 1990 imports from Japan of finished units, cars plus light trucks, declined by about 1 million units, and 2 million units of American-owned capacity were retired. At the same time, 2.5 million units of Japanese transplant capacity were added. Thus declining imports plus retired capacity exceeded new transplant capacity by 500,000 units and actual capacity utilization in 1990 was only slightly below the level of 1987, the drop-off being due to lower sales in 1990 compared with 1987.[19]

On another level, however, many problems still must be overcome if North America is to avoid the European fate of the 1920s, in which reforms in production were deferred for a generation. Many of these difficulties are internal to the production system itself, while others are political; some are both. They include:

- The cyclical pattern of the U.S. motor vehicle market, which is incompatible with lean production.
- North American notions of careers, which are also incompatible with lean production.
- The fact that the rapid decline of U.S. and Canadian-owned companies that many regard as national institutions is likely to prove too much for politicians and the general public to accept.

Let's take a closer look at each of these problems.

## LEAN PRODUCTION IN THE SEA OF CYCLICALITY

Westerners are resigned to the idea of the business cycle. Like gravity, it's simply there, although no one quite knows why. No one likes it, and cures have frequently been proposed, the most recent being Keynesian macroeconomic management. To date none has worked.

Mass production is, in fact, a system ideally suited to the survival of large enterprises in a highly cyclical economy. Both workers and suppliers are considered variable costs. When the market goes down, the assembler companies jettison their human and organizational ballast and expect to find their workers and suppliers pretty much where they left them once conditions improve. Figure 9.2 and Figure 9.3 show the pattern of demand and production in the United States over the past forty years.

**FIGURE 9.2**

**Cyclicality of the Motor Vehicle Market, United States Compared with Japan, 1946–1989**

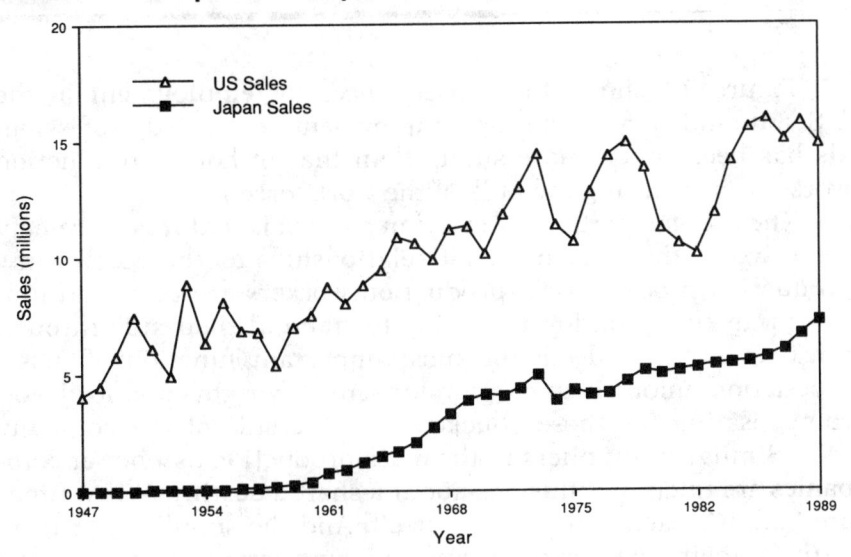

*Source:* United States sales: Motor Vehicle Manufacturers Association, *Motor Vehicle Facts and Figures*. Japanese sales: Japan Automobile Manufacturers Association, *Motor Vehicle Statistics of Japan*.

**FIGURE 9.3**

**Cyclicality of Motor Vehicle Production, United States Compared with Japan, 1946–1989**

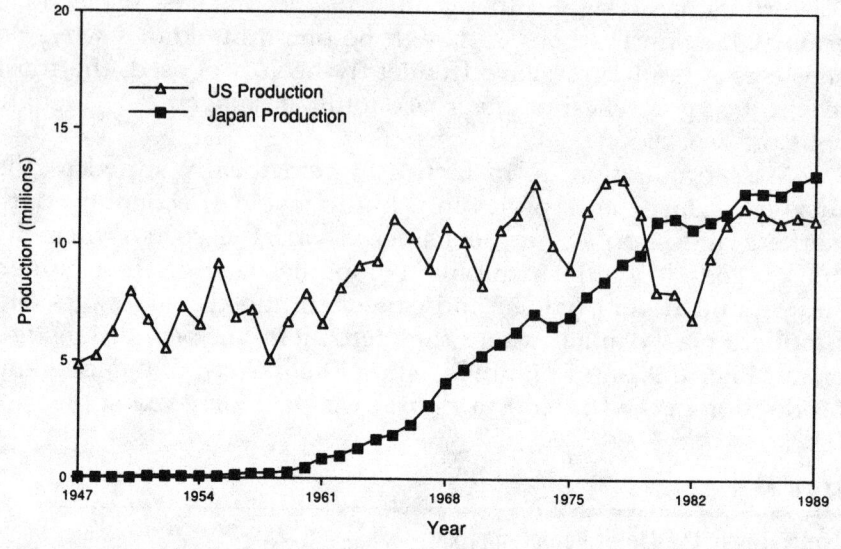

*Source: Automotive News Market Data Book*

Figure 9.4 shows the consequences for employment in the U.S. auto industry. (Note that employment of salaried professionals has been much more stable than that of hourly production workers constituting the bulk of the work force.)

The problem with the American pattern is that it is extremely corrosive to the vital personal relationships at the core of any production process. Mass-production workers are under no illusions that their employer is going to stand at their side through thick and thin. Indeed, the most important function of mass-production unions is to bargain for seniority rights and for layoff compensation for those chucked over the side of the company ship. Similarly, suppliers to the mass-production assembler companies are under no illusions about a shared destiny. When times are bad, it's each company for itself. And the suppliers, in turn, jettison their own workers and subcontractors. As noted, the consequence is a distinct lack of commitment on the part of workers and suppliers.

Lean production, by contrast, is inherently a system of recip-

## FIGURE 9.4

**Employment in the United States Motor Vehicle Industry, 1946–1989**

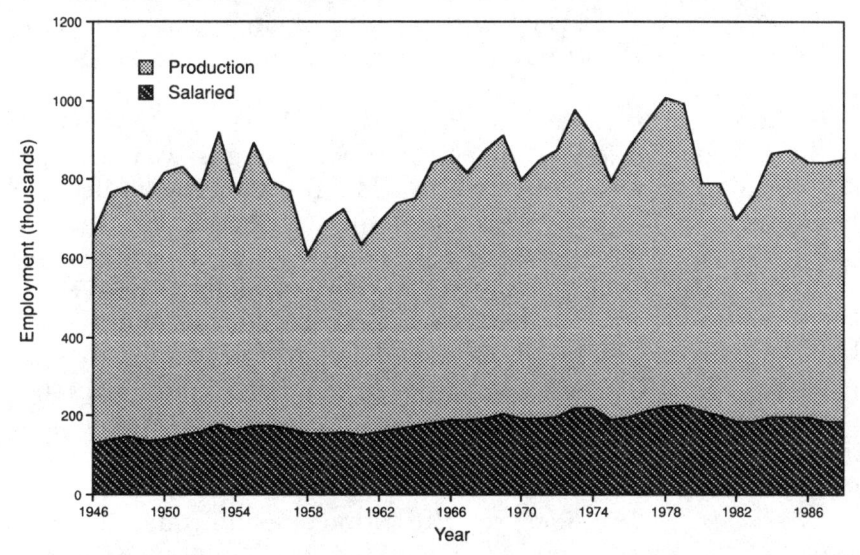

*Note:* The figure includes all jobs in Standard Industrial Classification 371, "motor vehicles and parts."

*Source:* United States Department of Labor, Bureau of Labor Statistics

rocal obligation. Workers share a fate with their employer, suppliers share a fate with the assembler. When the system works properly, it generates a willingness to participate actively and to initiate the continuous improvements that are at the very heart of leanness.

But can this system work in a cyclical economy? As figures 9.2 and 9.3 show, the issue never arose in Japan because neither the domestic auto market nor domestic production is cyclical. (As the low-cost, high-quality global producer of automobiles until very recently, the Japanese domestic industry has always been able to plow through slumps in export markets by cutting margins.) Indeed, the *largest* contraction in production in Japan over the past forty years is smaller than the smallest contraction in North America.

What happens as lean producers, whether Japanese or American, encounter heavy seas in North America (and, to a lesser degree, in Europe)? A General Motors executive gave us one answer, when he looked at a version of figures 9.2 and 9.3 during an interview: "When the Japanese [meaning lean] producers encounter these gigantic market waves, they will quickly become as mediocre as we are. They will have to start hiring and firing workers along with suppliers and will end up as mass-producers in short order."

We aren't so sure, but we do feel this is a vital issue to which hardly anyone in the West has given much thought: Management of the macroeconomy may have a dramatic long-term effect on the fundamental quality of the domestic production system. Those public officials who have periodically felt it necessary to break the back of inflationary expectations by deflating the economy may need to think anew about the probable consequences for the production system. If nascent lean producers act to save themselves in a steep slump by jettisoning their most valuable asset—their people—the real cost of poor Western macroeconomic management may prove even greater in the future than it has proved in the past.

More positively, widespread adoption of lean production may dampen both inflation and the business cycle. If mass production is ideally suited to the survival of big companies through deep cycles in demand, it may also be cycle-enhancing. That is, its penchant for massive inventories, both of in-process parts and finished units, would seem to exacerbate the cycle: As inflation builds, stocks are built up against expectations of yet higher prices. This move pushes prices up farther. Then, when the economy suddenly falters, the built-up stocks are worked off, deepening the slump upstream in the production system.

Some observers have even wondered if the lack of a cyclical market in durable goods in Japan is a direct result of lean production: An inventoryless, highly flexible system may significantly damp cyclicality.

The Japanese have another cycle damper in their arsenal in the form of flexible compensation. Most employees at all levels in Japanese companies receive a large part of their compensation—up to a third—in the form of bonuses directly tied to the profitability of the company. So when the market drops, at least in theory, the company can dramatically slash prices due to lower operating costs and restore production to its former level.

In fact, this system has been tested only at companies such as Mazda that have experienced a crisis independent of general market conditions. The simple reason is that there have been no deep cycles in the Japanese economy generally, so there has been no real test of employee tolerance for large wage cuts.[20]

## WESTERN "CAREERS" VERSUS JAPANESE "COMMUNITY"

This point leads directly to a second question confronting the future of lean production in the West. Why won't workers in companies temporarily cutting wages simply leave for better opportunities in other companies or industries? The answer in Japan is simple: By convention, practically all hiring in companies is from the bottom only and, as noted, compensation within the company is based largely on seniority. Jumping ship would be quite pointless, because the employee would almost always be worse off starting at the bottom elsewhere rather than waiting for the situation to improve with the current employer.

This situation obviously doesn't prevail in the West. In addition, as we pointed out in Chapter 5, Western notions of careers are quite different from the needs of lean production.

Most workers in the West place a very high value on having a portable skill—something they can take with them if things don't work out in a particular company. This concept is tied quite tightly to Western educational systems that stress discrete competences and certify students to prove that the skills have been attained. This concern about skills is very similar to the mind-set of the skilled craftsman, who was—and is—obsessed with maintaining his portable skills, although professional workers in the West rarely see the parallel.

However, as we saw, for the lean-production system to succeed, it needs dedicated generalists willing to learn many skills and apply them in a team setting. The problem, as we also noted in Chapter 5, is that brilliant team play qualifies workers for more and better play on the same team but makes it progressively harder to leave. So a danger exists that employees who feel trapped in lean organizations will hold back their knowledge or even actively sabotage the system. Western companies, if they are to become lean, will need to think far more carefully about personnel systems and career paths than we believe any have to date.

## THE POLITICS OF A PAINFUL TRANSITION

In our discussion so far, we've simply reported the obvious: Lean production is diffusing rapidly in North America but mostly under the leadership of Japanese companies. Yet, throughout human history, foreign investment and ownership seem always to operate on a razor's edge, continually testing the tolerance of host countries. Recent shifts in American attitudes confirm this point.

Initially the U.S. government was elated that the Japanese would be plowing money into U.S. automotive plants. At the same time, state governments fell over themselves with inducements to invest in one state rather than another. And, as the investments grew, the public came to accept the new plants as part of the landscape.

Recently, however, a new tone has emerged as the full logic of Japanese investment has begun to dawn on public officials, the U.S.-owned auto companies, the UAW, and U.S.-owned supplier companies.

First, the Japanese companies do not believe they can turn around existing mass-production facilities. Therefore all the transplants are "greenfields" in car talk, or entirely new plants, with the exception of NUMMI, which was a semi-greenfield, in the sense that it had been permanently closed by General Motors two years before it reopened under Toyota management.

Second, the Japanese aren't building the transplants simply to circumvent trade barriers or a temporarily strong yen. They've discovered that they can make cars in North America as well as they can in Japan and, even more important, that they can make cars in North America better than two of the three American-owned companies. It therefore follows that the transplants will keep on growing until the American companies improve their performance and regain the initiative, or are eliminated.

Third, it appears that the UAW won't be able to organize those transplants, whose owners have no links to the U.S. companies. The union was badly defeated in an election at Nissan in 1989 and, so far, has been unable to gather the petitions it needs for an election at Honda or Toyota. The Mazda, Diamond-Star (Mitsubishi-Chrysler), and CAMI (GM-Suzuki) plants have been organized, as has NUMMI, because, in each case, the plant has a connection to a unionized U.S. firm. However, most observers expect the plants of the Japanese Big Three (Toyota, Nissan, and

Honda) to grow the fastest. In consequence, the UAW has begun to wonder about its institutional future if nonunion transplants continue to displace the unionized plants of the U.S. Big Three.

If it fails to organize plants, the UAW's only option may be to seek a limit on their expansion through political means. Recently, the union has urged that vehicles assembled at the transplants be subtracted from the number of cars each Japanese assembler is allowed to import under the continuing Voluntary Restraint Agreement. The logical effect of such a policy would be to set permanent market-share limits on each Japanese company. This would seem to guarantee the survival of the unionized U.S.-owned companies.

Fourth, as the Japanese companies rapidly increase their domestic content (that is, the fraction of a car's value manufactured in the United States), U.S. suppliers are learning that it is not easy to supply the transplants, for reasons we examined in Chapter 6, and that it is hard to make much money doing so.

Putting all these facts together, we find it hardly surprising that members of the U.S. Congress, company executives, and union leaders are beginning to wonder if the unarguable success of the transplants is a cause for celebration or alarm. The North American motor industry stands every prospect of being revitalized to regain world-class performance during the 1990s. The massive trade deficit in motor vehicles is likely to shrink or even disappear. However, this lean machine may be largely foreign owned and nonunion if the American mass-producers don't improve their performance quickly.

We believe that the period through 1992 will prove the most tense. If GM and Chrysler fail to go through a creative crisis, one that breaks the logjam of old ideas and narrow interests and opens the path to lean production, and if the economy should slump badly during this period, we have great concerns about the outcome. We've shown just how plans can go awry when we discussed the European experience with mass production.

## THE TRANSITION TO LEAN PRODUCTION IN EUROPE

But what of lean production in the current stronghold of mass production? As we have seen, the European auto industry is today, after a fifty-year transition from craft production, the leading proponent of old-fashioned mass production—high volume, long product runs, infinitely fragmented work, "good enough" product quality, enormous inventories, massive factories. And, as we saw in the experience of the young French factory manager, there has been little pressure thus far to achieve leanness. But that pressure will emerge as the 1990s unfold.

First, the market may not be as vigorous as it was in the 1980s. Everyone can make money in a seller's market. And the opening of Eastern Europe may produce a sustainable boom for the 1990s. Then again, growing congestion, environmental problems, and approaching saturation in Western Europe may hold demand below recent levels. Only a small slacking off in demand will make a large difference in the profitability of the high-volume European firms, who depend on very high levels of capacity utilization compared with the Americans.

Second, the Americans will be applying what they've learned in a decade of desperate struggle in North America. Ford is already the most efficient producer in Europe, with the exception of a few of its English operations, such as Dagenham, that never really embraced mass production.

Third, and most important, the Japanese are arriving. In the 1980s, they were diverted to North America, a continent with no local-content rules to delay the start-up of assembly plants and a market where they had already captured a 22-percent share before the erection of trade barriers.

In Europe, by contrast, market limits in France (3 percent), Italy (about 1 percent), the United Kingdom (11 percent), and Spain (a 40-percent tariff on imported cars) had held the Japanese to an 11-percent share overall, and the rules of entry for manufacturing—invented quite amazingly by the "free-enterprise" Thatcher government in Britain—included a requirement that domestic content must reach 60 percent within two years of start-up and reach 80 percent a few years later.[21] These restrictions, in turn, have meant that Japanese companies can't just build assembly plants. They must construct an engine plant and develop local suppliers for a host of components simultaneously, greatly raising their start-up costs. Nevertheless, the Japanese are now rapidly making their presence felt, as previously shown in Figure 8.2.

Many in Europe have congratulated themselves on their aggressive stance toward Japanese investment. They view the U.S. approach—that is, practically free market access for any company willing to build an assembly plant adding 25 percent or even less of each car's total value—as extremely naive. (Cars made at the U.S. transplants, that is, don't count against the quota on Japanese finished-unit imports. And this holds true even if the assembler does nothing but screw together Japanese parts.) This approach yields only "screwdriver" plants, they argue, with very little manufacturing value added. The heart of the industry, they maintain, will remain in Japan.

In our view, it is not a matter of being hardheaded or naive but rather of understanding the internal logic of lean production. It is a system based fundamentally on doing as much manufacturing as possible at the point of final assembly. Once a lean producer starts down the path to assembly in a major regional market, the logic of the system tends powerfully to bring the complete complement of production activities, including product development, along as well. And sooner rather than later, as is happening in North America.

The actual effect of the European policies is to create a much more difficult transition in the 1990s. As they begin the process of adaptation, the European companies have sought to perfect mass production as the Japanese have continued to perfect leanness, and they are much farther behind than the Americans were in 1980—off the pace by a factor of three or four in many countries in terms of fundamental productivity.

Based on past experience, pressures will mount for a Fortress Europe policy—for example, permanent company market-share limitations no matter where the product is produced. In fact, such a policy was recently advocated by Peugeot president Jacques Calvet. To do so would freeze the current inefficiencies of the European producers and ensure that they slip even farther behind lean producers elsewhere, a disastrous outcome.

However, we expect a compromise to be reached in which the rate of Japanese advance is slowed but not halted by policies from Brussels, and in which the period of transition to lean production is stretched out well into the next century. For example, the European Economic Community (EEC) will probably establish an overall quota for Japanese imports to Europe and will establish some content requirements for assembly plants if their European-assembled vehicles are not to be subtracted from the import quota. Nevertheless, Japanese investments will be permitted in any European country, and the quota will be relaxed over time. So the European mass-production industry must eventually learn to compete with lean producers.

# COMPLETING THE TRANSITION

## 10

It took more than fifty years for mass production to spread across the world. Can lean production spread faster? Clearly we think it is in everyone's interest to introduce lean production everywhere as soon as possible, ideally within this decade.

In North America, the full implementation of lean production can eliminate the massive trade deficit in motor vehicles. We are certain that when there is no longer any difference between the North American and Japanese motor industries in productivity, product quality, and responsiveness to changing market demand, trade will more or less naturally come into balance.

In Europe, the home today of classic mass production, lean production can quickly triple the productivity of the motor-vehicle industry while providing more fulfilling jobs for factory workers, engineers, and middle managers. It can balance Europe's motor-vehicle trade as well.

In many of the developing countries, lean production is a means for rapidly developing world-class manufacturing skills without massive capital investments. These countries would need only to find markets for their new industrial capabilities, a point we will return to in a moment.

Truly, the case for moving quickly is overwhelming. In this concluding chapter, we offer some practical ideas on how the transition to lean production can be completed by the end of this century.

## THREE OBSTACLES IN THE PATH

### Obstacle 1: The Western Mass-Producers

The greatest obstacle in the path of a lean world is easy to identify: the resistance of the massive mass-production corporations that are left over from the previous era of world industry. These companies—General Motors, Renault, Volkswagen, Fiat—are so large and prominent in the industrial landscape of North America and Western Europe that no government can allow them to fail suddenly. Yet many have proved remarkably incapable of reforming their ways in the 1980s.

What's more, most of the conventional means governments use to aid their home-owned companies are counterproductive over the long term. We've already noted the perverse effect of the quotas negotiated by the U.S. and European governments at the beginning of this decade. While the quotas were useful in sending a signal to the Japanese that foreign manufacturing would be necessary in the long term (at a time when currencies were sending just the opposite message in the short term), they generated massive profits for the Japanese companies to use in financing their direct-investment campaigns in North America and Europe.

We think that it is neither practical nor desirable for these massive Western enterprises to be swept aside by Japanese lean producers, but they need more creative solutions than the kinds that have been conventionally proposed. These solutions must take several forms:

• First, every mass-producer needs a lean competitor located right across the road. We have found again and again that middle management and rank-and-file workers in a mass-production company begin to change only when they see a concrete, nearby example of lean production that can strip away all the cultural and economic explanations of why the other manufacturer is succeeding. U.S., Canadian, and English mass-production plants now all have a lean producer right across the road, but the continental European countries have lagged badly. Their relative performance on our productivity survey shows the result.

Even the Americans, Canadians, and English have only had examples of lean factory practice to examine up close. It is only now that lean research-and-development operations are being established nearby. Thus the North Americans and English have not made significant progress in adopting lean design, but we expect to see rapid improvement shortly. This trend is already occurring in supplier systems where many North American suppliers are learning better methods from their work with the transplants. They are then using this knowledge to improve their relationships with the U.S. assemblers. So improvements, which we'd normally expect to come from the top down—assemblers working to make their suppliers better—are, instead, coming from the bottom up.

- Second, mass-producers in the West need a better system of industrial finance, one that demands they do better while supplying the large sums that will be needed to turn these large companies around. Currently, most of the debate in the area of finance focuses on low-cost funds for Western companies and ways to dismantle the Japanese group system. Such proposals are no doubt well intentioned but miss the point: Giving mass-producers more money to spend on inefficient product development, inefficient factory operations, and more sophisticated equipment than they need is bound only to make things worse in the long run. And dismantling the Japanese group system would be to junk the most dynamic and efficient system of industrial finance the world has yet devised.

- Third, most mass-producers will need a crisis, what we have called a *creative crisis,* to truly change. Ford, as we saw, had such a crisis in 1982. The company heeled over so far that the senior executives on the bridge were practically catapulted into the raging torrent. The result: A company previously plagued with internal dissension as executives tried to advance their own careers and workers sought wages and benefits divorced from productivity gains suddenly had a new sense of purpose and team spirit in saving itself from oblivion. Organizational changes that had seemed impossible were suddenly easy. GM and the European mass-producers, by contrast, have had periods of low profits and crises from other causes but have never had the sense that their

## COMPLETING THE TRANSITION

whole approach to production is doomed. So they seem to be dying the death of a thousand cuts as the world's lean producers, now including Ford, steadily gain ground.

The trick for investors and bankers when the crisis comes is to offer help—but only in return for the company's realistic plans for achieving world-class performance by converting to lean production. Governments may also need to help out by establishing paid training programs for employees that companies can no longer use. These excess workers are the heart of the conversion problem. That's because workers at mass-production plants learn no skills. So when a mass-production company collapses, most workers can qualify only for the most entry-level jobs in other sectors. Training these displaced workers for meaningful work will be essential.

Indeed, a major problem in converting from mass to lean production is that in a highly competitive marketplace, where share growth is impractical, a substantial fraction of the work force is no longer needed. If the European volume producers were to convert to lean production today without gaining market share, they would need less than half their current work force. Doubtless, the market for cars and trucks would grow as competition pushed prices down, but it is unrealistic to think that shedding of labor can be avoided.

One of the most important tenets of the Toyota Production System is never to vary the work pace. Therefore, as efficiencies are introduced in the factory or design shop, or as the rate of production falls, it is vital to remove unneeded workers from the system so that the same intensity of work is maintained. Otherwise the challenge of continual improvement will be lost. The same is true of mass-production companies converting to lean production. Excess workers must be removed completely and quickly from the production system if improvement efforts are not to falter.

Companies such as GM have tried to do so by creating job banks of excess workers to be retrained for other jobs in the company. The problem is that, realistically, there will never be any other jobs within the mass-production companies and the ability of the companies to finance the job banks will decay over time as well. So some sort of public support for job banks may be necessary, and workers will have to be trained for positions outside the traditional manufacturing industries. This idea is sternly resisted by government officials and union leaders in many Western countries, the former because of the up-front cost to governments, the latter because the union loses strength as workers leave their companies. However, the alternative approach of propping up mass-producers through trade and investment barriers, so they can afford to continue their inefficient use of human effort, is far more costly in the long term.

### Obstacle 2: Outdated Thinking About the World Economy

Once upon a time, not very long ago, most people thought that the world economy advanced by moving the production of standardized, low-priced products—such as small automobiles and trucks—to new mass-production factories in newly industrializing countries. In the 1970s, the rise of Japan was often explained in this way.

Five years ago, when we started our project, many observers expected that Japan would soon find it could no longer compete in the export of small vans and trucks because of the strengthening of the yen and its effect on Japanese wages. Korea, Taiwan, Thailand, and Malaysia, the next tier of countries with low wages and an educated, industrious work force, would collectively become the next Japan. These countries, it was argued, would rapidly build up their economies the way Japan had—through the export of small cars and trucks to the United States and Europe, supplanting Japanese products in the process.

We never subscribed to this view, because we knew that lean production is more than a match for low-wage mass production. First, lean production dramatically raises the threshold of acceptable quality to a level that mass production, particularly in low-wage countries, cannot easily match. Second, lean production offers ever-expanding product variety and rapid responses to changing consumer tastes, something low-wage mass production finds hard to counter except through ever lower prices. Continually dropping prices is unlikely to work, however, because a third advantage of lean production is that it dramatically lowers the amount of high-wage effort needed to produce a product of a given description, and it keeps reducing it through continuous incremental improvement, as we saw in Chapter 4. Finally, lean production can fully utilize automation in ways mass production cannot, further reducing the advantage of low wages. The expansion of the Korean motor industry in the 1980s sums up this situation.

In 1979, Korea was nowhere in the auto industry. Despite government protection of the domestic market since 1962, the Korean industry, consisting of four small producers, had not gotten very far. Hyundai was the most advanced and, unlike its rivals, was to a considerable degree independent of the U.S., European, and Japanese assemblers. Its Pony model used an independently engineered body and an engine and transmission wholly manufactured in Korea from designs licensed from Mitsubishi. It had enjoyed some modest export success, mostly in third-world markets, such as in Latin America, where it sold on price. The other small companies—Daewoo, Kia, and Dong A—made vehicles solely for sale in the Korean domestic market, using designs licensed from European companies. Unlike Hyundai, they were completely dependent on their European partners for technology.

The world economic crisis of 1979 and 1980 hit particularly hard in Korea. Domestic sales collapsed. So did Hyundai's exports, once Japanese companies dropped their prices in order to maintain their sales in export markets. This crisis gave the Korean Ministry of Industry the chance it had been waiting for—the opportunity to rationalize the industry over the opposition of the *chaebol* (Korea's version of the Japanese *keiretsu*) in the way the Ministry of International Trade and Industry (MITI) had wanted to do in Japan in the 1950s.

The Ministry pushed Kia and Dong A out of the industry for five years, while assigning Hyundai small cars and Daewoo larger ones. Hyundai, in particular, took this directive as a sign to proceed down the path to high-volume mass production. It began to plan a new model, the Excel, which would be built in a massive new factory in Ulsan, mostly for export to the United States and Europe. The Excel was almost entirely based on licenses from Mitsubishi in Japan. Indeed, it was practically indistinguishable in general specifications from Mitsubishi's Colt model. Hyundai's strategy was simple: It would compete by underpricing the Japanese entry-level cars, based on low wages and high volume.

For a brief period, the strategy worked brilliantly. Hyundai's Excel arrived in the vital U.S. market in 1986, just as the Japanese were raising their prices to counter the strengthening of the yen. Americans assumed that any Asian car, particularly one with a Japanese design, would have Japanese quality. And, at a price

$1,000 below Japanese vehicles in the same size range, the Excel seemed unbeatable. Sales quickly grew to 350,000 per year, and Hyundai scrambled to add capacity in Korea with a second 300,000-unit assembly plant.

The Korean Ministry of Industry was so impressed with Hyundai's triumph that it soon permitted Kia back into the auto market. Kia would build a small car based on the Mazda 121 to be sold in the United States by Ford as the Festiva. In addition, Daewoo was allowed to build a second, smaller model, based on the German Opel Kadett, for sale in the United States through General Motors dealers as the Pontiac LeMans. By 1988, the Korean producers were selling 500,000 cars in the United States, accounting for 4 percent of the total market.

Then the Korean strategy fell apart. Hyundai was, in fact, an old-fashioned mass-producer, with low wages but a large number of hours expended per car. When the Korean currency began to strengthen rapidly against the dollar in 1988, and Korean auto workers demanded large wage increases, a large part of the Koreans' cost advantage disappeared. At that point, the question of quality emerged. The early Hyundai cars sold in the United States had very poor quality, as shown in the J. D. Power data we used in our assembly-plant studies in Chapter 4.

In 1987, when the average Japanese car was reported by consumers to have about 0.6 defects, the Hyundai cars had 3.1. As the word began to spread, the Korean producers found it necessary to slash prices to sustain sales, just at the point when their production costs were soaring. The consequence was that Korean sales in the United States *fell 50 percent between 1988 and 1990.* The next Japan was no longer the next Japan.[1]

What's more, by the late 1980s, it had become apparent that there would be no next Japan, even if a developing country created a lean-production industry that could match the product quality and labor productivity of the best lean producers. Japan's success had so sensitized the world trading system to massive inflows of industrial products from one region to another that no country could realistically hope to pick up where Japan left off. Indeed, at the peak of the Korean auto sales in North America in 1988, the U.S. government pushed the Korean government hard to reduce its growing overall trade surplus by 50 percent, which the Koreans did.

Hyundai became convinced that to protect its access to the

North American market, it would need to follow the Japanese transplants in building an assembly plant in North America. Its 100,000-unit Bromont, Quebec, plant opened in 1989 to assemble the mid-size Sonata, a new model it hoped would restore its fortunes in the North American market. The idea of a company from a developing country building a major manufacturing facility in a highly developed, high-wage country would have been unthinkable only five years earlier, at a time when most observers predicted an inexorable drift of low-tech manufacturing, including cheap autos, out of the developed world. After all, Korea's entire advantage was expected to be low wages. But because Korean wages are rapidly converging with those in North America, and political considerations require manufacture within the North American region, Korea is now assembling cars in Canada.

What do we conclude from the experience of Hyundai and the other Korean producers? That the world economy, in a short period, has changed in remarkable ways. First, the triumph of lean production has created a new threshold for product quality that no producer can hope to offset merely through low prices based on low wages. As a result, producers in the next tier of developing countries must become lean producers as well. As we will see shortly, it is quite feasible for them to move toward their goal in the 1990s.

Second, even those developing countries mastering lean production will need to think anew about the market for their products. Partly, they should look at home, because the productivity gains of lean production should bring motor vehicles into reach of a much larger fraction of domestic consumers. In Brazil, for example, we found that fifty hours were needed to assemble our standard small car *compared with thirteen* at the best Japanese lean producer. Not surprisingly, the Brazilian car market has been stuck at about 1 million units for many years. About one-third of the difference in performance is due to greater automation in the highest-productivity Japanese plant, but complete introduction of lean production without advanced technology should cut the level of effort in Brazil in half, opening a vast new domestic market.

The developing countries should also look for regional markets. Indeed, the most striking feature of the world economy in the past few years is the sudden reorientation of the trading patterns in manufactured goods from cross-regional, across the great oceans, to intraregional, within the great regions—North America, Europe, East Asia.

The motor industry is perhaps the leader in this trend. Exports from Japan to Europe are stable, while Japanese and European exports to North America are falling dramatically and European exports to Japan are growing dramatically from a very small base. (We don't, of course, count as exports cars manufactured in the United States and Europe by Japanese companies.) What we expect by the end of the decade is a much lower volume of total exports between regions, a greater balance in the remaining trade flows, and a focus for the remaining cross-regional trade on niche-type, specialty products. This approach, of course, is precisely the one proposed for the post-national lean enterprises we described in Chapter 8.

Meanwhile, within the great regions, the flow of products among countries should increase dramatically. Let's begin with North America. The United States and Canada began to integrate their auto industries in 1965, when the Canada–U.S. Auto Pact went into effect. For the participating assembler companies, the U.S. Big Three, this meant that cars and trucks could be made in one country and shipped to the other for sale without paying tariffs—as long as the assemblers met modest Canadian requirements to keep Canadian production roughly proportional to Canadian sales. (This point quickly became moot, as Canada began to run a substantial trade surplus with the United States.) In 1989, the Canada-U.S. Free Trade Agreement set in motion the final process of integration of the automotive market by eliminating by the mid-1990s all remaining tariffs on the flow of parts between the two countries.

An interesting issue in the North American region is Mexico. For a period of thirty years, from the beginning of the 1960s, Mexico tried to develop a domestic motor industry that could supply the Mexican market with all its needs. To achieve this goal, the Mexican government in 1962 prohibited the imports of finished vehicles and imposed high local-content requirements on the five foreign companies—Ford, GM, Chrysler, Nissan, and Volkswagen—building cars in Mexico.[2]

The policy was both a spectacular success and a spectacular failure. By 1980, Mexico had a 500,000-unit motor-vehicle industry producing vehicles with perhaps 50-percent local content.

## COMPLETING THE TRANSITION

Unfortunately, the industry—with market-share restrictions and a host of other protections for the domestic assemblers and parts suppliers—was totally uncompetitive in both cost and quality in the world market. With five producers making three or four separate models each in a 500,000-unit market, average annual production totaled about 25,000 units of each product, far too low even for today's lean producers to make economically. What's more, the Mexican plants were in no way lean. Even Nissan, a lean producer in Japan, employed a combination of craft and mass-production methods at its Cuernavaca plant.

Mexican policy might have continued on its course except for the economic collapse beginning in 1981. As domestic demand tumbled and Mexico's foreign debt mounted in 1983, the government examined its auto policy. Its initial strategy was to push even farther down the path to mass production by limiting each assembler to a single product while raising the required domestic content level. Government officials reasoned that although Mexican car and truck buyers would have very limited choices, economies of scale should cause the cost of cars and trucks to fall as volume increased. What was more, the Mexican trade deficit in automotive products should fall as domestic content grew.

It soon become apparent that this strategy wouldn't work. The domestic market was simply too small and the protected domestic producers too inefficient. Mexico would need to join the world. The first step was to permit Ford to build a new assembly plant in the northern city of Hermosillo. This plant had no domestic content requirement as long as it exported the great majority of its output.

The Hermosillo plant also provided the first opportunity to experiment with lean production in Mexico. At this greenfield site, Ford applied what it had learned from Mazda in building a Mazda-designed car, sold in the United States as the Mercury Tracer. Hermosillo was a great success in terms of productivity and quality. Mexican workers embraced lean production with the same speed as American workers at the Japanese transplants in North America and at Ford's own U.S. and Canadian plants. However, the plant failed to meet its cost targets, because it was assembling its cars entirely from parts shipped from Japan. As the yen strengthened, Hermosillo, a plant envisioned by Mazda and Ford in the early 1980s as a way around the U.S.-instigated quota on Japanese finished-unit imports, suddenly made no sense. What does make sense—and is also consistent with the concept of lean production—is to manufacture a large fraction of the parts—engine, transmission, and so forth—for the car at Hermosillo and serve the entire North American regional market, including Mexico.

The Mexican government dramatically altered its strategy at the end of 1989 to make this approach more feasible, not just for Hermosillo but for the entire Mexican motor-vehicle industry. It dramatically reduced its local-content requirements on individual products and relaxed its rules against the import of finished cars and trucks while retaining its requirement that companies making and selling cars in Mexico balance their trade by exporting as much as they import.[3]

With this step, a new configuration of production for the entire North American region could emerge. GM, Ford, Chrysler, Nissan, and VW might assemble in Mexico—for sale to the entire North American market—cheap, entry-level cars and trucks that use parts produced by production complexes in northern Mexico, near the assembly plants. At the same time, larger cars and trucks for Mexican consumers might be supplied by plants in the U.S. and Canadian Midwest. While Mexico would run a substantial trade surplus with the United States and Canada, this integration of Mexico into the North American region should actually be a net gain for the U.S. and Canadian auto industries. That's because exports to Mexico from U.S.-Canadian plants would be additional business (no finished-vehicle shipments have been permitted for the past thirty years). What's more, the Mexican market might grow very rapidly from its current base of 500,000 units to 2 million or more by the year 2000. The small cars and trucks manufactured in Mexico for sale in the United States and Canada would logically replace imports from Japan, Korea, and Brazil. Currently, these imported small vehicles provide no jobs in the United States and Canada.

For this logic to be fulfilled, one change in U.S. policy will be necessary. The U.S. government's fuel-economy regulations will need to be modified so that small cars produced in Mexico by U.S. companies, with high levels of Mexican manufacturing value, are treated as "domestic products." Otherwise, the U.S. companies will not be able to participate fully in Mexico, leaving its potential to be realized by Japanese or European companies that are not similarly constrained. And some way must be found around the 25-percent American tariff on pickup trucks. The Mexican government proposal in March 1990 to open negotiations on a North

American free trade zone may provide the best means of addressing this problem.

A similar process of regional integration is now expected in Europe in the 1990s. The initial impetus was the European Community's decision to remove remaining barriers to the flow of goods within the Community beginning in 1993. This move motivated the countries of the European Free Trade Area (Norway, Sweden, Iceland, Austria, Switzerland, and Finland) to seek their own integration into the European market. These historic decisions have been overshadowed for the moment by the dramatic changes in Eastern Europe and in the Soviet Union, changes that suddenly raise the prospect of a gigantic European market of up to 750 million consumers (including those in Russia and the other European republics of the Soviet Union). If this market should actually come to pass, it would be three times the size of the U.S.-Canadian markets and seven times larger than Japan's.

For the motor-vehicle industry, the logic of a united European region is similar to that for an integrated Mexican-Canadian-U.S. market. We expect that Eastern Europe will replace Spain as the production locale for the most inexpensive, basic cars and trucks and that the growing economies of Hungary, Czechoslovakia, Poland, and East Germany, in particular, will provide a market for larger cars and trucks produced in Western Europe. For example, Volkswagen has just begun assembly of its smallest model, the Polo, at the Trabant factory in East Germany. It plans to increase production to 250,000 units by 1994. GM has also formed a joint venture with the other East German producer, Wartburg, to produce 150,000 Opel Kadetts. Fiat, meanwhile, has announced major ventures in Poland and the Soviet Union for production of up to 900,000 vehicles, many of which may be sold in Western Europe badged as Fiats.

As in the case of Mexico, we also expect the East European countries to run a surplus in motor-vehicle trade. Like Mexico, they are all deeply in debt and can scarcely afford to run a further deficit in automobiles, which are still consumer luxuries in most of these countries. Just as in the North American region, the West European auto industry can actually gain production volume from a fully integrated Europe if the low-cost entry-level products produced in East Europe for sale in the West displace imports from East Asia.

East Asia is itself the third emerging region, although behind

North America and Europe in its development. Only a few years ago, the individual economies of Japan, Korea, and Taiwan were struggling to boost their exports of finished manufactured goods for the North American and European markets. They seemed almost oblivious of each other and were highly resistant to accepting manufactured goods imported from their neighbors. This situation is now changing rapidly, due partly to trade barriers and currency shifts that close off markets in the other regions, partly in response to the movement toward regionalization in Europe and North America. In 1989, for the first time since World War II, trade within the East Asian region exceeded extraregional trade with North America and Europe.

The logic of motor-vehicle industry development in East Asia is similar, save in one respect, to that of North America and Europe. We expect more basic vehicles to be made in top-to-bottom manufacturing complexes in the developing countries of the region for sale in all countries of the region. We also expect the manufacture of more complex and expensive vehicles to be focused in Japan for export to other markets in the region. Indeed, this trend is already beginning. Hyundai, Kia, and Daewoo all plan to begin selling entry-level vehicles in Japan in 1991. The Korean domestic market, thus far closed to finished Japanese vehicles, will be opened a crack at the same time. While the domestic Japanese industry, unlike those in Western Europe and North America, is unlikely to prove better off under this arrangement, it can easily be no worse off if its exports of more luxurious vehicles grow enough to offset decreased domestic production of basic vehicles.

The anomaly in East Asia, of course, is China. Until spring 1989, it seemed to be moving toward a more open stance regarding its economy and the world and might logically have entered into a regional East Asia market on at least a limited basis. Perhaps it can yet do so in the 1990s, but for the moment the Chinese industry is still focused inward, pursuing a combination of extremely rigid mass production in its two volume-production complexes in Changchun (No. 1 Auto Works) and Hubai (No. 2 Auto Works) and inefficient low-quality craft production in about a hundred additional vehicle-manufacturing facilities spread throughout China.[4]

This disastrous combination gives China the distinction of having the world's largest motor-vehicle industry in terms of employment (more than 1.6 million workers) and one of the

## COMPLETING THE TRANSITION

smallest in terms of output (a projected 600,000 units in 1990). By contrast, in 1989, 500,000 employees in the Japanese auto industry produced 13 million vehicles, indicating a productivity gap of about seventy to one between two countries separated by a hundred miles across the Sea of Japan.

So much for the three great regions comprising about 90 percent of the current-day motor-vehicle market. What of those countries, such as Brazil and Australia, with substantial motor industries, and others such as India, with substantial aspirations? Where do they fit into the emerging world of regions and regional production systems? Our belief is that they must look primarily to their own regions for markets, but in creative ways. Let's take the two very different examples of Brazil and Australia.

Brazil set out in the late 1950s to build a top-to-bottom motor-vehicle production system. It permitted the multinational car companies, notably GM, Ford, Volkswagen, and Fiat, to own 100 percent of the equity in their Brazilian operations but insisted that they quickly convert from kit building, using imported parts, to the use of practically 100-percent Brazilian parts in each vehicle. By the mid-1960s, in the midst of the Brazilian economic miracle, this goal had been achieved. The Brazilian industry reached 1 million units of production annually.[5]

Unfortunately, for twenty years now, Brazil has been a story of stagnation. As we noted, the mass-production complexes built in Brazil were a notable achievement compared with the alternative—complete dependence on imports. However, these plants now lag far behind the world pace in terms of productivity and product quality. In addition, in the early 1970s, after oil prices soared, the government required that the industry introduce alcohol-fueled engines, a requirement that focused the industry's product-development energies on a technology that has found no market elsewhere in the world. Meanwhile, the number of years each model was kept in production soared to fourteen years in Brazil, nearly four times the Japanese standard.

For a brief period in the mid-1980s, the Brazilian industry thought it had found a new strategy: It would take advantage of its low wages by exporting cheap cars to Europe and the United States. (The models in question were the Volkswagen Fox subcompact sold in the United States and the Fiat Duna sold in Europe.) This was a Latin variant of the Korean strategy, which met with the same pattern of high hopes, based on initial sales, followed by despair as currencies shifted and product shortcomings offset an initial price advantage. Fox sales in the U.S. market, for example, fell from a peak of 60,000 in 1987 to 40,000 in 1989. GM, meanwhile, canceled a tentative plan for Brazilian production of a mini-mini van, based on its German Opel Kadett, mainly for export to the United States.

A more promising path for Brazil in the 1990s will consist of three elements. First, lean producers must show Brazil the way toward world-class manufacturing. Honda's motorcycle plant at Manaus, far up the Amazon, has clearly demonstrated that lean production can work in Brazil under the most demanding conditions, but automotive examples in the Brazilian industrial heartland near São Paolo are essential.[6] Introduction of lean production can dramatically reduce production costs to spur the stagnant domestic market, where only the upper-middle class can now afford the output of the inefficient mass-production auto industry.

Second, Brazil needs to open its industry to imports of whole vehicles and parts, so that real competition will be introduced into what is now a tight oligopoly. Because Brazil can hardly afford to run a trade deficit in motor vehicles, given its massive foreign debt, it will no doubt need to require that producers balance their trade. However, a truly competitive market can still be achieved with a flexible policy. The new Mexican Auto Decree shows one way to accomplish this end.

Third, Brazil will need to integrate its production system with its neighbors, beginning with Argentina.[7] By setting in motion a regionalization process as it brings production costs down, Brazil can develop a massive Latin American growth market that will not depend on favorable trade policies and currencies in the other great regions. While productive trade with those regions should still be possible, it will not be the key to the strategy. Thus Brazil and its neighbors can control their own fate.

Australia presents perhaps the most difficult instance of a country with a small and highly developed motor-vehicle industry but with an insufficient domestic market and, thus far, a lack of regional outlook. The Australian government decided in the 1960s that it would develop a top-to-bottom auto industry to replace both imported vehicles and the kit assembly of parts produced in Europe and North America. By the end of the 1960s it had done so but with all the disadvantages of mass production in a low-volume, highly protected market. Despite Australia's efforts in the 1980s to consolidate the five producers into three more viable production systems and the presence of several Japanese producers, our IMVP assembly-plant surveys have found productivity and quality levels far off the standard set by lean producers in Japan and North America.

For a while in the mid-1980s, Australia thought that perhaps it too could succeed by following Korea's example. Ford proposed to export a specialty vehicle, a Mazda 323 modified as a convertible roadster, to be sold in the United States under the name Mercury Capri. This was at a time when the Australian dollar was very weak and the American dollar very strong. However, by the time the car was ready for production and a number of quality problems had been resolved, currencies had shifted and the car no longer made much economic sense.[8] The effort illustrates once more the risk of an extraregional export strategy in a world of fluctuating currencies.

The logical path for Australia would be to reorient its industry toward the Oceanic regional market including Indonesia, Singapore, and the Philippines. Each country within this region might balance its motor-vehicle trade, but, collectively, by permitting cross-shipment of finished units and parts, they could gain the scale needed to reduce costs and let lean production flourish. Australia, as the most advanced country in the region, presumably would concentrate its own production on complex luxury vehicles, while Indonesia at the other extreme, would make cheap, entry-level products.

Unfortunately, nothing of this sort has happened. Australia views itself as part of the developed world and thinks naturally of exporting to North America, Europe, and even Japan, while Indonesia thinks of itself as part of the developing world of the Association of South East Asian Nations (ASEAN) countries and focuses on developing trade with Malaysia, the Philippines, and Thailand. Repeated efforts to develop an ASEAN-mobile by pooling parts produced by different companies in each country have come to nothing, because it doesn't make sense to do so in terms of the commercial strategy of the multinational assembler and components firms.

Thus the Oceanic countries of the Southern Hemisphere constitute a region still waiting to happen. The same can be said of the countries of the Indian subcontinent and those of southern

Africa. As the rest of the world pursues a path of regionalization in the 1990s, we expect that regional thinking will grow in these areas as well. The combination of regional scale and lean production can be a particularly powerful stimulus to growth if the right policies are followed.

### Obstacle 3: Inward Focus of the Japanese Lean Producers

The final obstacle to a lean world is, in fact, the Japanese lean producers themselves. How can this be? Many of you will, no doubt, have concluded that we think everything these companies do is good, compared with the bad practices of the Western mass-producers. In one way this impression is accurate: These companies have provided an invaluable gift to the world by pioneering a new way of making things that really is superior. But in another way they lack a final and essential innovation: the ability to think and act globally rather than from a narrow national perspective.

Anyone who reads the newspapers is aware of the growing backlash to Japanese direct investment in North America and Europe, what the Japanese themselves call investment friction. We regard this trend as a much greater threat to the eventual creation of a lean world than trade barriers on finished units and parts could ever be. This is because, in the worst case, it can lead to investment barriers that permanently seal off North America, Europe, and the other regions from Japanese lean competitors that can force everyone to become lean.

Why is this backlash building when the Japanese companies are creating new jobs in new manufacturing complexes making cars, trucks, and parts at levels of quality and productivity equal to those of the home plants in Japan? Partly it stems from the threat these facilities represent to established institutions—mass-production companies and mass-production unions. Friction, for these reasons, is an inescapable component of change and progress.

However, there is another, more fundamental reason for friction. Many government officials, managers, and workers in the West perceive that the Japanese lean producers are offering two classes of citizenship in their organizations—one for Japanese workers, a second for foreigners; one for Japanese suppliers, another for foreign suppliers; and one for Japanese group mem-

## COMPLETING THE TRANSITION

bers of their *keiretsu*, but none at all for foreign companies. As Westerners watch the seemingly inexorable advance of the Japanese companies, this second-class citizenship begins to seem unacceptable. As one manager at GM remarked, "I can hope to get to the top at GM, but I can never hope to rise above the middle level of one of the Japanese foreign subsidiaries, no matter how superior my performance." Growing investment friction is the result. The outcome is uncertain.

Executives at the Japanese companies are acutely aware of this problem and have given it much thought. One solution, now being pursued by several of the auto companies, is to appoint native managers to head their manufacturing operations in North America and Europe. Similarly, a number of Japanese companies are designating native supplier companies as their source for certain categories of components. Governments in both regions are supporting this approach through restrictions on visas for Japanese employees at the new facilities and, in Europe, through strong pressures to attain high levels of domestic content as soon as possible. (The latter policy raises the cost and launch time of the initial facilities substantially unless most parts are obtained from existing domestic suppliers.)

The consequence, we fear, will be a repetition of Ford's experience in Britain after 1915. The wholesale substitution of domestic managers and suppliers, to deal with investment friction, quickly degraded the performance of Ford's production system toward the existing English level. While Ford did spur English producers to adopt new ways, the full benefits of mass production were never achieved.

This is not an idle fear, based on events long ago. In our assembly-plant survey of the transplants in North America and Europe, we found strong evidence that those plants that perform best are those with a very strong Japanese management presence in the early years of operations, and those that have moved slowly and methodically to build up their domestic supply base. The performance of other plants, which have turned over most management to North Americans and Europeans recruited at senior levels from Western car makers, and who have hurriedly assembled a supply team, is better than the Western average but in many cases not as good as the Western company—Ford—that has taken lean production to heart.

It should be clear that the "Japaneseness" of the management and the suppliers is not the issue. Rather, it is how well transplant managers and suppliers understand lean production and how deeply they are committed to making it work. Unfortunately, at the moment, a large fraction of the world's managers with knowledge and commitment to leanness are Japanese.

We would suggest that a better approach for the Japanese companies will be to build a truly global personnel system in which new workers from North America, Europe, and every other region where a company has design, engineering, and production facilities, are hired in at an early age and given the skills, including language skills and exposure to management in different regions, needed to become full citizens of these companies. This will mean an equal opportunity to head the company someday.

Similarly, the Japanese lean assemblers will need to form supplier groups in each region where they operate by exchanging shares in supplier firms and offering full citizenship. They will also need to regionalize their equity base and borrowings so that shifting currencies will not hinder the most appropriate deployment of production in each region. Finally, a truly important advance in terms of its visibility will be for the *keiretsu* to include foreign companies in their membership. For example, those *keiretsu*, such as that of the Dai-Ichi Kangyo Bank, with weaker auto companies among their members (Isuzu and Suzuki in this instance) might invite a strong Western car company to join. On the other hand, the one Japanese car company unaffiliated with a *keiretsu*, Honda, might wish to form an international *keiretsu* consisting of Western manufacturing companies and a bank.

For any of these innovations to work, a clear two-way understanding will be essential. Western companies and employees will need to embrace the concept of reciprocal obligation, making a long-term commitment to the company or the group. Japanese companies, in turn, will need to abandon their narrow national perspective and quickly learn to treat foreigners who accept the obligations involved as full citizens.

We are well aware how difficult these innovations will be to implement. American and European companies have struggled, in many cases for decades, to provide full citizenship for foreigners in their organizations. Yet there are still no foreigners in the senior management of General Motors or on its board, and it made headlines recently when Volkswagen appointed Daniel Goudevert, a Frenchman, as the first foreigner on its management board.

In addition, the Japanese will need to address issues of ethnic background and gender in providing citizenship for foreign employees, issues that they have not confronted in Japan (where there are practically no minorities and where women are notably absent from senior management) and where Japanese practice is far behind Western norms.

Nevertheless, the Japanese must logically be the innovators in devising postnational, multiregional corporate forms and providing full citizenship to their employees and suppliers drawn from many countries and regions across the globe. They have the financial resources that many Western companies lack and they have the need; they risk investment barriers and other impediments to the expansion of their production systems if they don't.

They should start right away by declaring their intention to proceed as postnationals (companies where nationality is not related to promotion prospects) and by implementing postnational personnel, supplier, finance, and *keiretsu* systems that the outside world can examine.

"Transparency"—the ability for outsiders to see the system in action, understand its logic, and verify its performance—is critical to Western acceptance, because of the long time lag between initiating such a system and proving that it really works (for example, as young employees entering at the bottom rise to the top). One highly visible way to demonstrate their intent would be for Japanese companies to assign newly hired Westerners to work for several years in Japan, where there are presently practically no permanent non-Japanese employees of the big companies.

Only a public and emphatic commitment to these final organizational innovations—which Western firms must match as well—will ensure the triumph of lean production, for the Japanese companies and for the whole world. Such a commitment will also provide part of the essential glue to hold together the emerging world regions in North America, Europe, and Asia—regions no longer united by the familiar East/West conflict and in danger of drifting apart in the twenty-first century.

# EPILOGUE

When Henry Ford and Alfred Sloan created mass production, the ideas they incorporated were in the air all about them. Everywhere there was a sense that the older craft-based modes of production had reached their limits. What was more, many parts of the mass-production system had been tested previously in other industries. The meat packing industry, for example, had pioneered moving "disassembly" lines for cutting up carcasses before the turn of the century. In the 1890s, the bicycle industry had pioneered many of the steel stamping techniques and dedicated machine tools Ford later used. Even earlier, the transcontinental railroads had developed many of the organizational mechanisms for managing large firms operating over vast areas.

But Ford and Sloan were the first to perfect the entire system—plant operations, supplier coordination, management of the entire enterprise—and to couple it with a new conception of the market and a new distribution system. Thus, the auto industry became the global symbol of mass production.

The complete system spread rapidly to other industries in the United States in the 1920s and was soon embraced by practically all volume manufacturing industries. In addition, mass production was tried, without much success, in one-of-a-kind craft industries—in particular housing, where a number of entrepreneurs set out to become the Henry Ford of the home.

In Europe, the idea of mass production was a problem not just for the auto industry, but in every industry. On one level, the intellectuals, particularly on the left, embraced the idea of mass production as the obvious means to elevate the living conditions of the masses. Soon the images of mass production and modernity were a central theme of European art. However, back at the factory, in every type of manufacture, the poor fit between the requirements of mass production and the craft orientation of both workers and managers insured that adoption of the new techniques was very slow. The lack of an integrated European market was a further impediment. It was only after World War II that mass production was fully embraced across the industrial landscape of Europe, in many cases through use of "guest workers" from other countries and cultures who were willing to tolerate the monotony of classic mass production in the factory.

Just as Ford and Sloan were swimming in the sea of new ideas, the postwar chaos in Japan created a fertile environment for new thinking. Many of the techniques Eiji Toyoda and Taiichi Ohno built into their lean production system were being tried at the same time in other industries. For example, the quality-enhancing ideas of the American consultant W. Edwards Deming were adopted at about the same time by many Japanese companies across a range of industries. A number of other ideas were forced on these inventors by larger social forces in society, in particular the need to treat workers as fixed costs once it became apparent that hire-and-fire labor policies would be strenuously resisted by employees.

However, like Ford and Sloan, their achievement lay in putting all the pieces together to create the complete system of lean production, extending from product planning through all the steps of manufacture and supply system coordination on to the customer. Thus, the auto industry once more changed the world and has become the global symbol of the new era of lean production.

What's more, as we have seen, lean production combines the best features of both craft production and mass production—the ability to reduce costs per unit and dramatically improve quality while at the same time providing an ever wider range of products and ever more challenging work. The final limits of the system are not yet known and its diffusion, both within the auto industry and to other industries, is still at an early state—about where mass production was in the early 1920s. Yet in the end, we believe, lean production will supplant both mass production and the remaining outposts of craft production in all areas of industrial endeavor to become the standard global production system of the twenty-first century. That world will be a very different, and a much better, place.

# End Notes

### BEFORE YOU BEGIN THIS BOOK

1. Alan Altshuler, Martin Anderson, Daniel Jones, Daniel Roos, and James Womack, *The Future of the Automobile,* Cambridge: MIT Press, 1984.

### CHAPTER 1

1. Peter Drucker, *The Concept of the Corporation,* New York: John Day, 1946.
2. See, for example, Ford Motor Company Chairman Harold Poling's speech to the *Automotive News* World Congress, January 7, 1990, in which he estimated that worldwide "excess capacity" in the motor vehicle industry would reach 8.4 million units in 1990.
3. For an excellent assessment of General Motors' predicament, see Maryann Keller, *Rude Awakening: The Rise, Fall and Struggle for Recovery at General Motors,* New York: William Morrow, 1989.

### CHAPTER 2

1. The material in this section on Evelyn Ellis and his car was obtained in the archives of the Science Museum, London. It consists of newspaper accounts of Ellis's exploits and an internal background memo prepared by the museum staff on the 1894 Panhard car belonging to the museum.
2. The material on Panhard et Levassor is from James Laux, *In First Gear: The French Auto Industry to 1914,* Liverpool: Liverpool University Press, 1976.
3. Ford acquired majority control of Aston Martin in 1987. It also acquired the small British sports car builder AC in that year. Other craft producers acquired by multinational auto firms in the 1980s were Lotus (General Motors), Ferrari (Fiat), and Lamborghini (Chrysler).

4. Ford proposed this term in his 1926 article for the *Encyclopedia Britannica*, "Mass Production" (13th edition, Suppl. Vol. 2, pp. 821–823). Many others at the time called his techniques "Fordism."
5. Two extraordinarily useful studies of mass production in the factory, as pioneered by Ford, are David Hounshell, *From the American System to Mass Production, 1800–1932*, Baltimore: Johns Hopkins University Press, 1984, particularly chapters 6 and 7, and Wayne Lewchuk, *American Technology and the British Vehicle Industry*, Cambridge: Cambridge University Press, 1987, particularly Chapter 3. The account given here of the origins of the Ford system is taken from these sources, unless otherwise indicated.
6. In 1919 the entire vehicle assembly department at Highland Park employed only $3,490 of capital equipment (Lewchuk, *American Technology*, p. 49).
7. Ford's ability to cut prices during the life of the Model T is summarized by William Abernathy, *The Productivity Dilemma: Roadblock to Innovation in the Automobile Industry*, Baltimore: Johns Hopkins University Press, 1978, p. 33.
8. *The Ford Manual*, Detroit: Ford Motor Company (no date), pp. 13, 14.
9. This survey is cited in Daniel Raff, "Wage Determination Theory and the Five-Dollar Day at Ford," Ph.D. dissertation, Massachusetts Institute of Technology, 1987, an interesting study of the social implications of Ford's system.
10. This oversight may help explain why the productivity of the whole factory did not improve at the same rate as that of the assembly lines. See Lewchuk, pp. 49–50.
11. Alfred D. Chandler, *The Visible Hand: The Managerial Revolution in American Business*, Cambridge: Harvard University Press, 1977.
12. This information and subsequent material on Ford's organization and operations are taken from Allan Nevins and Frank Ernest Hill, *Ford: The Times, the Man, the Company*, New York: Scribners, 1954; Allan Nevins and Frank Ernest Hill, *Ford: Expansion and Challenge, 1915–1932*, New York: Scribner's, 1957; and Mira Wilkens and Frank Ernest Hill, *American Enterprise Abroad: Ford on Six Continents*, Detroit: Wayne State University Press, 1964. The specific information on U.S. assembly plants is from Nevins and Hill, *Ford: Expansion and Challenge*, p. 256. The figure for foreign assembly plants is derived from Wilkens and Hill, Appendix 2.
13. Alfred P. Sloan, *My Years with General Motors*, Garden City, New York: Doubleday, 1963. Peter Drucker presented his own codification in *The Concept of the Corporation* in 1946. Henry Ford II read this volume when taking over from his grandfather that year and set out to remake Ford in GM's image.
14. For the best explanation of the logic of mass-production unionism see Harry Katz, *Shifting Gears: Changing Labor Relations in the U.S. Automobile Industry*, Cambridge: MIT Press, 1985.

## CHAPTER 3

1. *Toyota: A History of the First 50 Years*, Toyota City: Toyota Motor Corporation, 1988, provides a useful summary of Toyota's history.

# END NOTES

2. The Toyota production total has been calculated from *Toyota: A History*, p. 491. Toyota had also produced 129,584 trucks between 1937 and 1950, mostly for military use. The Rouge production figure includes 700 vehicles assembled at the Rouge and 6,300 kits of parts Ford shipped to its final assembly plants spread across the United States.
3. *Toyota: The First 30 Years*, Tokyo: Toyota Motor Company, 1967, pp. 327–328 (in Japanese).
4. In the interest of brevity we have skipped over the many conceptual contributions of the Toyota Motor Corporation's founding genius, Kiichiro Toyoda. Kiichiro Toyoda had a number of brilliant insights in the 1930s, inspired in part by his own visit to Ford in Detroit in 1929. These included the just-in-time supply coordination system. However, the chaotic conditions in Japan in the 1930s prevented him from implementing most of his ideas.
5. For an excellent summary of the development of the Toyota Motor Corporation and the techniques of lean production, see Michael Cusumano, *The Japanese Automobile Industry: Technology and Management at Nissan and Toyota*, Cambridge: Harvard University Press, 1985.
6. Toshihiro Nishiguchi, "Strategic Dualism: An Alternative in Industrial Societies," Ph.D. dissertation, Nuffield College, Oxford University, 1989, pp. 87–90, provides a good analysis of the consequences of the new labor laws imposed by the American occupation. One of the many ironies of Japanese-American relations is that both a new approach to labor relations and a new system of industrial finance were imposed on Japan by American occupation officials sympathetic to President Franklin Roosevelt's "New Deal," who had been unable to gain the political support for similar measures in the United States. Two of Roosevelt's most vehement and effective opponents in the area of labor law reform were Alfred Sloan and Henry Ford.
7. Toyota and the other auto companies did employ considerable numbers of temporary workers for many years as they struggled to keep up with growing demand and resisted granting life-time employee status to workers they were not sure they could retain. However, this practice came to an end in the 1970s as the Japanese firms gained confidence that their growth was not an accident and could be sustained.
8. The Introduction to Michael Cusumano's, *Japanese Automobile Industry* provides a succinct account of MITI's twenty-year effort to reorganize the industry and its failure to do so.
9. Interested readers are urged to consult Ohno's work directly for the details of his innovations: Taiichi Ohno, *The Toyota Production System*, Tokyo: Diamond, 1978 (in Japanese). An excellent account in English, prepared with Ohno's help, is Yasuhiro Monden, *The Toyota Production System*, Atlanta: Institute of Industrial Engineers, 1983.
10. As we will see in Chapter 6, a key additional problem of this system lay in devising a workable bookkeeping system so that the real production cost of in-house parts operations was known. To outside suppliers, it often appeared that arbitrary allocation of corporate overheads made make/buy decisions sham proceedings rigged in favor of the in-house supplier.
11. Ohno and Monden provide detailed explanations of this system in their volumes on the Toyota Production System.
12. We define a model as a vehicle with entirely different exterior sheet metal from other products in a producer's range.

13. For details on their efforts see Shotaro Kamiya, *My Life with Toyota*, Tokyo: Toyota Motor Sales Company, 1976.
14. Toyota Motor Sales was created during the crisis of 1949 at the insistence of Toyota's bankers. They believed that a separate sales company would be less likely to produce over-optimistic sales forecasts leading to excessive production than the previous system, in which marketing was simply another division of the Toyota Motor Company. Certainly, the trauma of a vast inventory of unsold products in 1949 spurred Toyota's thinking on how to build the inventoryless system that eventually emerged. Toyota Motor Sales was remerged with the Toyota Motor Company in the late 1980s to form the current Toyota Motor Corporation.

# CHAPTER 4

1. This was a major change since Satoshi Kamata's slashing critique of working conditions at Toyota in the early 1970s (*Japan in the Passing Lane: An Insider's Account of Life in a Japanese Auto Factory*, New York: Pantheon, 1982 (originally published in Japan in 1973). In the early 1960s more than 40 percent of Toyota's work force was temporary workers without permanent job guarantees. By 1975 all temporary workers had been converted to permanent workers, a situation that continued until 1989, as Toyota strained to keep up with the burst of auto demand in Japan and once more hired workers without permanent guarantees. We will return to the problems that demand fluctuations create for lean production in Chapter 9.
2. Throughout the program and throughout this volume we have used information on product quality provided by J. D. Power and Associates, an American firm specializing in consumer evaluations of motor vehicles. However, we do not use the "Power numbers" now routinely cited in automobile advertising in North America. These numbers are for defects in the entire vehicle. Because we have been interested in the activities of only one part of the manufacturing system, the assembly plant, we have obtained data from Power on defects that can be directly attributed to the activities of the assembly plant. Specifically, these are water leaks, loose electrical connections, paint blemishes, sheet metal damage, misaligned exterior and interior parts, and squeaks and rattles.

    Because the Power data are only available for vehicles sold in the United States, the number of European, Japanese, and New Entrant plants for which we can report quality data is smaller than the number for which we have productivity data and other indicators of manufacturing performance.
3. This is the method used by many of the publicly available comparisons of productivity in auto industry. See, for example, Harbour Associates, *A Decade Later: Competitive Assessment of the North American Automotive Industry, 1979–1989*, 1989.
4. For a full explanation of our methods the reader should consult John Krafcik, "A Methodology for Assembly Plant Performance Determination," IMVP Working Paper, October 1988.
5. We are pledged not to reveal the identity of specific plants and, by logical extension, of companies. However, the dramatic improvement in Ford's plant-level performance in the 1980s is now so well known that it seems unrealistic not to acknowledge it.

6. The advantage of having suppliers across the road who can deliver high-quality parts directly to the line every hour or two is considerable. In the American transplants, where most parts are delivered much less frequently, considerable effort still goes into inspecting incoming parts and then transferring them to the point on the line where they are installed.
7. This finding is being borne out by a number of studies in other industries as well. See, for example, Joseph Tidd, "Next Steps in Assembly Automation," IMVP Working Paper, May 1989, for a comparison of recent experience with automation in the automotive and electronics industries, and R. Jaikumar, "Post Industrial Manufacturing," *Harvard Business Review*, November/December 1986, pp. 69–76, for a study of flexible automation in machine shops and the watch industry.
8. For details of this survey see John Krafcik, "The Effect of Design Manufacturability on Productivity and Quality: An Update of the IMVP Assembly Plant Survey," IMVP Working Paper, January 1990.
9. For details on model-mix complexity and under-the-skin complexity as predictors of assembly plant productivity and quality, see John Krafcik and John Paul MacDuffie, "Explaining High Performance Manufacturing: The International Automotive Assembly Plant Study," IMVP Working Paper, May 1989.
10. We do not mean to suggest that Ford has no plans to eventually renegotiate its rigid job-control contracts. The contract at the Wayne, Michigan, assembly plant was recently renegotiated in the direction of a team concept as a prerequisite to Ford's decision to allocate the new Ford Escort to the plant.
11. See Mike Parker and Jane Slaughter, "Managing by Stress: The Dark Side of the Team Concept," in *ILR Report*, Fall 1988, pp. 19–23, and Parker and Slaughter, *Choosing Sides: Unions and the Team Concept*, Boston: South End Press, 1988.

## CHAPTER 5

1. The Clark team's findings are reported in the following:

    Kim B. Clark, W. Bruce Chew, and Takahiro Fujimoto, "Product Development in the World Auto Industry," Brookings Papers on Economic Activity, No. 3, 1987.

    Takahiro Fujimoto, "Organizations for Effective Product Development: The Case of the Global Automobile Industry," Ph.D. thesis, Harvard Business School, 1989.

    Kim B. Clark and Takahiro Fujimoto, "The European Model of Product Development: Challenge and Opportunity," IMVP Working Paper, May 1988.

    Kim B. Clark and Takahiro Fujimoto, "Overlapping Problem-Solving in Product Development," in K. Ferdows, *Managing International Manufacturing*, Amsterdam: North Holland, 1989.

    Kim B. Clark and Takahiro Fujimoto, "Product Development and Competitiveness," paper presented at the OECD Seminar on Science, Technology and Economic Growth, Paris, June 1989.
2. Takahiro Fujimoto, "Organization for Effective Product Development," tables 7.4 and 7.8.
3. Their findings are summarized in:

    Antony Sheriff, "Product Development in the Auto Industry: Corporate Strategies and Project Performance," master's thesis, Sloan School of Management, MIT, 1988.

Kentaro Nobeoka, "Strategy of Japanese Automobile Manufacturers: A Comparison Between Honda Motor Company and Mazda Motor Corporation," master's thesis, Sloan School of Management, MIT, 1988.
4. Clark and Fujimoto, "Product Development in the World Auto Industry," p. 755.
5. This example is based on material presented in Clark and Fujimoto, "Overlapping Problem-Solving in Product Development."
6. Clark and Fujimoto, "Overlapping Problem-Solving in Product Development," Table 2.
7. Clark and Fujimoto, "Product Development in the World Auto Industry," p. 765. In addition, although Clark and Fujimoto do not report data on this, the downtime during which a plant is not running at all during a model changeover is much briefer in lean-production plants.
8. Earlier results plus the methodology used are reported in:

Antony Sheriff, "The Competitive Product Position of Automobile Manufacturers: Performance and Strategy," IMVP Working Paper, May 1988.

Antony Sheriff and Takahiro Fujimoto, "Consistent Patterns in Automotive Product Strategy, Product Development, and Manufacturing Performance," IMVP Working Paper, May 1989.
9. Alfred P. Sloan, *My Years with General Motors*, Garden City, New York: Doubleday, 1963, p. 72.
10. This dilemma is one of those spelled out in William Abernathy, *The Productivity Dilemma: Roadblock to Innovation in the Auto Industry*, Baltimore: Johns Hopkins University Press, 1978.
11. The material in this section is based on Andrew Graves, "Comparative Trends in Automotive R&D," IMVP Working Paper, May 1987.
12. These figures update earlier data reported in Andrew Graves, "Comparative Trends in Automotive R&D," and Daniel Jones, "Measuring Technological Advantage in the World Motor Vehicle Industry," IMVP Working Paper, May 1988.
13. For information on the European and American efforts in this area see Andrew Graves, "Prometheus: A New Departure in Automotive R&D," IMVP Working Paper, May 1988, and Hans Klein, "Towards a U.S. National Program in Intelligent Vehicle/Highway Systems," IMVP Working Paper, May 1989.
14. Perhaps the most interesting, among the vast outpouring of literature on the greenhouse effect, is James Lovelock, *The Ages of Gaia: A Biography of Our Living Earth*, New York: Norton, 1988.

## CHAPTER 6

1. Toshihiro Nishiguchi, "Competing Systems of Automotive Components Supply: An Examination of the Japanese 'Clustered Control' Model and the 'Alps' Structure," IMVP Working Paper, May 1987.
2. There are really two issues confounded in the claim that "lower supplier costs" are the Japanese competitive advantage. One is the number of hours of effort needed to accomplish a set of activities. As we will see, there is good reason to think that Japanese suppliers require much less

effort, just as Japanese assemblers need much less effort to design cars and to assemble the parts. The second issue is the cost per hour of effort. At one time in Japan there was a large gap between wages in the assembler firms and the lower levels of the supplier system. However, as Toshihiro Nishiguchi has recently shown ("Strategic Dualism," Ph.D. thesis, Oxford University, 1989, pp. 155–156), this gap shrank by the 1960s to only about 20 percent, which happens to be about the gap between assembler and supplier wages on average in the United States today. For an assembler that is very highly vertically integrated—General Motors is the obvious example—this gap does still produce a cost gap with competitors—such as Chrysler—obtaining a large fraction of their parts from outside suppliers.

3. Richard Lamming, "The International Automotive Components Industry: Customer-Supplier Relationships, Past, Present, and Future," IMVP Working Paper, May 1987, provides a good historical overview of the changing relations between assemblers and suppliers in North America and Europe.

4. Richard Lamming remembers the situation at Jaguar a decade ago when he was told to justify his job by "saving his salary," that is, by finding immediate cost savings on purchased parts sufficient to cover his cost to the company. In addition, it was commonly understood that the next promotion would go to the purchasing agent producing the largest savings, perhaps two or three times his salary. Until recently this approach was typical in the purchasing departments of Western companies. It served to institutionalize a system squeezing prices down in the short term at the expense both of costs and assembler-supplier relationships in the long term.

5. Much of the material in this section is based on the work of IMVP research affiliate Toshihiro Nishiguchi:

"Competing Systems of Automotive Component Supply," IMVP Working Paper, May 1987.

"Reforming Automotive Purchasing: Lessons for Europe," IMVP Working Paper, May 1988.

"Strategic Dualism," Ph.D. dissertation, Oxford University, 1989.

6. Richard Lamming, "The Causes and Effects of Structural Change in the European Automotive Components Industry," IMVP Working Paper, May 1989, pp. 22–23.

7. Toshihiro Nishiguchi, "Strategic Dualism," p. 210.

8. For a full description of this system see Nishiguchi, "Strategic Dualism," p. 191.

9. The authors are grateful to Richard Hervey of Sigma Associates for bringing this point to their attention.

10. Nishiguchi, "Strategic Dualism," p. 202.

11. For numerous examples of assemblers sharing ups and downs with suppliers, see Nishiguchi, "Strategic Dualism," pp. 281—311.

12. See Nishiguchi, "Strategic Dualism," pp. 281–311, for specific examples.

13. Access to that part of the plant making the part in question. It is important to remember that most suppliers in Japan do work for more than one assembler and, often, for firms outside the auto industry as well. Activities for other assemblers are off limits, since the supplier must maintain a close and confidential relationship with those firms as well.

14. Konosuke Odaka, Keinosuke Ono, Fumihiko Adachi, "The Automobile Industry in Japan: A Study of Ancillary Firm Development," Oxford: Oxford University Press, and Tokyo: Kinokuniya, 1988, pp. 316–317.
15. Nishiguchi, "Strategic Dualism," pp. 203–206, provides numerous examples.
16. It is important to note that actual practice differs considerably between assemblers. Toyota designates two or more suppliers for most smaller parts—for the front disk brake calipers for the base model Corolla, for example. Nissan and Honda, by contrast, maintain several suppliers for a given category of parts—for example front brake calipers in general. However, Nissan and Honda do not dual- or triple-source a specific part—the caliper for the front brakes of a specific model. Instead, they assign one specific part to each supplier and then compare their overall performance. If a supplier lets down on a specific part, it is relatively easy to transfer some of the business to another supplier of that type of part. So in practice, the Nissan and Honda systems are functional equivalents of the Toyota system.
17. Nishiguchi, "Competing Systems of Automotive Components Supply."
18. Takahiro Fujimoto, "Organizations for Effective Product Development," Table 7.1 Also see, Figure 6.3 in this chapter.
19. Nishiguchi, "Competing Systems of Automotive Components Supply," p. 15.
20. His findings are reported in Richard Lamming, "Structural Options for the European Automotive Components Supplier Industry," IMVP Working Paper, May 1988; "The Causes and Effects of Structural Changes in the European Automotive Components Industry," IMVP Working Paper, 1989; and "The International Automotive Components Industry: The Next Best Practice for Suppliers," IMVP Working Paper, May 1989.
21. This survey was conducted by Susan Helper of the Boston University School of Management. She has reported her findings in "Supplier Relations at a Crossroads: Results of Survey Research in the U.S. Automobile Industry," Boston University School of Management Working Paper 89-26, 1989.
22. Richard Lamming, "Causes and Effects of Structural Change in the European Automotive Components Industry," pp. 22–23.
23. Toshihiro Nishiguchi, "Strategic Dualism," p. 197.
24. Susan Helper, "Supplier Relations at a Crossroads," p. 7.
25. Helper, "Supplier Relations," p. 12.
26. Nishiguchi, "Strategic Dualism," pp. 116, 203, 204.
27. Helper, "Supplier Relations," Figure 7.
28. Nishiguchi, "Strategic Dualism," p. 218.
29. Helper, "Supplier Relations," Figure 7.
30. Helper, "Supplier Relations," p. 7.
31. Nishiguchi, "Strategic Dualism," pp. 313–347, and Nishiguchi, "Is JIT Really JIT?" IMVP Working Paper, May 1989.
32. John Krafcik, "Learning from NUMMI," IMVP Working Paper, September 1986; and Nishiguchi, "Strategic Dualism," p. 213.
33. Krafcik, "Learning from NUMMI," p. 32.
34. This section is based on Richard Lamming, "Causes and Effects of Structural Changes in the European Automotive Components Industry."
35. Takahiro Fujimoto, "Organizations for Effective Product Development," Table 7.1.

36. Lamming, "Causes and Effects," p. 39.
37. Lamming, "Causes and Effects," p. 43.

## CHAPTER 7

1. *Automotive News Market Data Book*, various years.
2. John J. Ferron, "NADA's Look Ahead: Project 2000," IMVP Working Paper, May 1988.
3. *Automotive News Market Data Book*, various years.
4. Calculated by the authors from *Automotive News Market Databook*, 1989 edition, p. 38.
5. John J. Ferron and Jonathan Brown, "The Future of Car Retailing," IMVP Working Paper, May 1989, and Jonathan Brown, "What Will Happen to the Corner Garage?" Brighton Polytechnic, Inaugural Lecture, 26 June 1988.
6. SRI International, *The Future for Car Dealerships in Europe: Evolution or Revolution?*, Croydon, U.K.: SRI International, July 1986.
7. Ferron and Brown, "The Future of Car Retailing," p. 11.
8. Data supplied by Professor Garel Rhys of Cardiff Business School.
9. This section draws on the work of Professor Koichi Shimokawa of Hosei University in Tokyo and on a case study of the Toyota distribution system prepared by Jan Helling of Saab.
10. *The Automobile Industry: Japan and Toyota*, published by Toyota Motor Corporation, Tokyo.
11. Koichi Shimokawa, "The Study on Automotive Sales, Service and Distribution Systems and Its Further Revolution," IMVP Working Paper, May 1987.
12. Koichi Shimokawa, "The Study of Automotive Sales," p. 30; *Automotive News Market Data Book*, and Japan Automobile Manufacturers Association, *Motor Vehicle Statistics of Japan*, Tokyo: JAMA, 1989.
13. Ferron and Brown, "The Future of Car Retailing," pp. 4–5.
14. John J. Ferron and Jonathan Brown, "The Future of Car Retailing," IMVP Working Paper, 1989, p. 11, and private communication with Koichi Shimokawa.
15. Other necessary changes have been the recent elimination of special excise taxes on cars with large engines (which penalized larger imported cars of the type with the most inherent sales appeal in Japan) and the willingness of several Japanese assemblers, under substantial international pressure, to retail imports through their own distribution channels. For example, Honda has recently begun selling Rover products in Japan through its Verno channel.

## CHAPTER 8

1. Allan Nevins and Frank Ernest Hill, *Ford: Decline and Rebirth*, New York: Scribner's, 1963.
2. We are indebted to Maryann N. Keller of Furman Selz Mager Dietz and Birney for an explanation of the Japanese system of capital formation in the 1980s and for the specific figures cited.
3. For an excellent summary of the Ford Motor Company's foreign operations between 1905 and the early 1960s, see Mira Wilkens and Frank

Ernest Hill, *American Business Abroad: Ford on Six Continents*, Detroit: Wayne State University Press, 1964. Unless otherwise indicated, the information on Ford's foreign operations cited here and in Chapter 9 is from this source.
4. Martin Adeney, *The Motor Makers: The Turbulent History of Britain's Car Industry*, London: Collins, 1988, p. 216.
5. Indeed, Honda was already exceeding Chrysler's passenger car sales in early 1990. However, it is important to realize that Chrysler also makes large numbers of mini-vans and trucks, so Honda has some ways to go in passing Chrysler in overall production.

## CHAPTER 9

1. Lawrence Seltzer, *A Financial History of the American Automobile Industry*, New York, 1928.
2. The material in this section is based on Wayne Lewchuk, *American Technology and the British Car Industry*, Cambridge: Cambridge University Press, 1988.
3. Lewchuk, *American Technology*, p. 153.
4. Lewchuk, *American Technology*, p. 155.
5. Lewchuk, *American Technology*, p. 157.
6. Mira Wilkens and Frank Ernest Hill, *American Industry Abroad: Ford on Six Continents*, Detroit: Wayne State University Press, 1964.
7. Thomas Husher, *American Genesis: A Century of Invention and Technological Enthusiasm*, New York: Penguin Books, 1989, p. 474.
8. This photo is reproduced in David Hounshell, *From the American System to Mass Production, 1800–1932*, Baltimore: Johns Hopkins University Press, 1984, p. 320.
9. Quoted in Lewchuk, *American Technology*, p. 175.
10. Quoted in Lewchuk, *American Technology*, p. 176.
11. This term is Lewchuk's invention.
12. The continental Europeans had also far surpassed the British industry, which never fully implemented mass production (until the 1980s) and sank rapidly before the competitive assault of the German, French, and Italian industries.
13. This material is from "Cost of Building a Comparable Small Car in the U.S. and in Japan—Summary of Consultant's Report to the UAW." This is a summary of a more detailed study prepared by Ford and the UAW with the help of consultants, which remains confidential.
14. This quote is from Maryann N. Keller, *Rude Awakening*, pp. 87–88.
15. The Japanese companies are limited to 11 percent of the market in Britain, 3 percent in France, 2,000 cars per year in Italy, and must enter Spain over a 40 percent tariff. In addition, the other "free trade" countries, led by Germany and Sweden, have periodically indicated that their continued support of an open market for Japanese cars is predicated on "appropriate" behavior by the Japanese firms, meaning in practice that their market share should only grow very slowly and without serious threat to the home country producers.
16. Chrysler acquired Rootes (British) and Simca (French) in 1963/64, amalgamating them into Chrysler Europe. This firm was sold to PSA in 1978.
17. J. J. Servan Schreiber, *The American Challenge*, New York: Atheneum, 1968.

END NOTES

18. GM opened its new Saturn complex at Spring Hill, Tennessee, in mid-1990. This is the company's single and very ambitious effort to implement the full set of lean techniques for product development, supplier coordination, and factory operations at a "greenfield" site where old ways of thinking can be cast aside. Saturn opened too late in the IMVP to be evaluated for this volume.
19. A complete enumeration of assembly plants in North America would add the opening of Hyundai's 100,000-unit assembly plant at Bromont, Quebec, in 1989 and Volkswagen's closing of its 250,000-unit Westmoreland, Pennsylvania, assembly plant, also in 1989. (This plant was subsequently sold to Sony for conversion to production of television picture tubes.) These adjustments would reduce net North American assembly capacity by another 150,000 units.
20. Entire industrial sectors in Japan have suffered prolonged slumps due to changes in the world economy. Steel and shipbuilding are the two prominent examples of the 1980s. When such slumps occur, Japanese government and industry exhibit a remarkable ability for restructuring and rationalization through the mechanism of the "recession cartel," in which excess capacity is retired in an orderly manner and the financial pain is shared equally among industry participants. Typically, rather than permanent pay cuts, capacity is retired and excess workers are transferred to growing companies within the *keiretsu*. However, this situation has never arisen in the auto industry.
21. The British government unofficially explained that this was necessary to guarantee that the cars would not count against Japanese import quotas when shipped to France and Italy. However, the government also recognized that Rover's and Ford's British production might decline as the Japanese transplants in Britain came on stream and was concerned about the fate of the British parts industry.

## CHAPTER 10

1. The reader should not jump to the conclusion that the Korean industry is "finished" in international competition. Although the initial quality of the Hyundai products was poor, their cars have improved steadily in the Power ratings to about average by 1989. This indicates that the Korean companies have both the ability and desire to learn from their errors and to make rapid improvements. In addition, the Korean firms have rapidly improved the sophistication of their factories. One firm has introduced many lean production techniques and needed only 25.7 hours of assembly effort in 1989 to perform our standard activities on the standard car, a level very near the Japanese average when the considerably lower level of Korean automation is factored in. Thus the real question marks for the Koreans are whether they can gain technological independence of the Japanese and Americans and whether they can find a stable role in the emerging East Asian region. Regarding the former see Young-Suk Hyun, "A Technology Strategy for the Korean Motor Industry," IMVP Working Paper, May 1989.
2. For a review of the Mexican situation see James P. Womack, "The Mexican Motor Industry: Strategies for the 1990s," IMVP Working Paper, May 1989.

3. See the "Mexican Auto Decree" promulgated by the government December 19, 1989, in *Diario Official*.
4. For a review of the Chinese situation, see Qiang Xue, "The Chinese Motor Industry: Challenges for the 1990s," IMVP Working Paper, May 1989.
5. For a review of the Brazilian situation, see Jose Ferro, "Strategic Alternatives for the Brazilian Motor Vehicle Industry in the 1990s," IMVP Working Paper, May 1989.
6. IMVP Research Affiliate Jose Ferro visited the Honda motorcycle plant at Manaus, far up the Amazon near the Peruvian border, and was amazed at the degree to which Honda had been able to implement lean production using rural migrants with absolutely no previous industrial experience. This is surely the most difficult environment in which lean production has thus far been tested and argues strongly that the basic ideas are truly universal.
7. This idea has been formally accepted by Brazil and Argentina, but to date the chaos in both economies has delayed serious progress toward implementation. For a review of the Argentine situation see Javiar Cardozo, "The Argentine Automotive Industry: Some Unavoidable Issues for a Re-entry Strategy," IMVP Working Paper, May 1989.
8. This car finally reached the American market in mid-1990.

## EPILOGUE

1. For a review of the impact of mass production on European thinking, see Thomas Hughes, *American Genesis: A Century of Invention and Technological Enthusiasm*, New York: Penguin Books, 1989, particularly chapter 6, "Taylorismus + Fordismus = Amerikanismus," and chapter 7, "The Second Discovery of America."

# APPENDIX A

## INTERNATIONAL MOTOR VEHICLE PROGRAM SPONSORING ORGANIZATIONS

AKZO nz
Australia—Department of Industry, Technology and Commerce
Automotive Industry Authority of Australia
Canada—Department of Regional Industrial Expansion
Chrysler Motors Corporation
Commission of the European Communities
Committee of Common Market Automobile Constructors
Daimler-Benz AG
Du Pont de Nemours & Co. Automotive Products
Fiat Auto SpA
Ford Motor Company
General Motors Corporation
Japan Automobile Manufacturers Association
Japan Automotive Parts Industry Association
Mexican Association of the Automobile Industry
Mexican Autoparts National Industry Association
Montedison Automotive Corporate Group
Motor and Equipment Manufacturers Association
Motorola, Inc.
Ontario—Ministry of Industry, Trade and Technology
Peugeot, SA
Quebec—Ministry of Industry and Commerce

Regie Nationale des Usines Renault
Robert Bosch GmbH
Rover Group
Saab Car Division
Swedish National Board for Technical Development
Taiwan—Ministry of Economic Affairs
TRW Automotive
United Kingdom—Department of Trade and Industry
United Kingdom—Economic and Social Research Council
United States—Department of Commerce
United States—Department of Transportation/NHTSA
United States—Office of Technology Assessment
Volkswagen AG
Volvo Car Corporation

# APPENDIX B

## INTERNATIONAL MOTOR VEHICLE PROGRAM RESEARCH AFFILIATE TEAM

Caren Addis, MIT
Jonathan Brown, Brighton Business School—United Kingdom
Javier Cardozo, University of Sussex—United Kingdom
Matts Carlsson, Chalmers University of Technology—Sweden
Al Chen, MIT
Joel Clark, MIT
Kim Clark, Harvard Business School
Michael Cusumano, MIT
Dennis DesRosiers, DesRosiers Automotive Research—Canada
Jose Roberto Ferro, Universidad Federal de Sao Carlos—Brazil
John Ferron, JD Power and Associates
Frank Field, MIT
Takahiro Fujimoto, Harvard Business School
Lars-Erik Gadde, Chalmers University of Technology—Sweden
Andrew Graves, University of Sussex—United Kingdom
Susan Helper, Boston University
Gary Herrigel, MIT
John Heywood, MIT
Young-suk Hyun, Han Nam University—Korea
Masayoshi Ikeda, Chuo University—Japan
Daniel Jones, Cardiff Business School—United Kingdom
Trevor Jones, MIT Senior Advisor
Christer Karlsson, European Institute for Advanced Studies in Management—Belgium

Harry Katz, Cornell University
Hans Klein, MIT
Thomas Kochan, MIT
John Krafcik, MIT
Donald Kress, MIT Senior Advisor
Richard Lamming, Brighton Business School—United Kingdom
Richard Locke, MIT
John Paul MacDuffie, MIT
Dennis Marler, MIT
Lars-Gunnar Mattsson, Stockholm School of Economics—Sweden
Noah Meltz, University of Toronto—Canada
Gian Federico Micheletti, Politecnico di Torino—Italy
Roger Miller, University of Quebec—Canada
Toshihiro Nishiguchi, MIT
Kentaro Nobeoka, MIT
John O'Donnell, MIT
Taku Oshima, Osaka City University—Japan
David Ragone, MIT Senior Advisor
David Robertson, MIT
Daniel Roos, MIT
Charles Sabel, MIT
Shoichiro Sei, Kanto Gakuin University—Japan
Luba Shamrakova, MIT
Antony Sheriff, MIT
Haruo Shimada, Keio University—Japan
Koichi Shimokawa, Hosei University—Japan
Joseph Tidd, University of Sussex—United Kingdom
Konomi Tomisawa, Long-Term Credit Bank—Japan
Kung Wang, National Central University—Taiwan
James Womack, MIT
Victor Wong, MIT
Qiang Xue, MIT

# APPENDIX C

## IMVP Program and Forum Participants*

UMBERTO AGNELLI—Chairman, Fiat Auto SpA
SHOICHI AMEMIYA—Director General, Nissan Mexicana, S.A. de C.V., Mexico
JOHN BANIGAN—Director General, Automotive, Marine and Rail Branch (FAMR), Department of Regional Industrial Expansion, Government of Canada
THEODORE BARDOR—President of Board of Directors, TEBO, S.A. de C.V. and TEBO Group, Mexico
T. R. BEAMISH—Chairman, The Woodbridge Group, Canada
FERNAND BRAUN—Director General for Internal Market and Industrial Affairs, Commission of the European Communities
GIANCARLO BERETTA—Director, Automotive Corporate Group, Montedison
STEPHEN BOWEN—Assistant Secretary, Department of Trade and Industry, U.K.
MICHAEL CALLAGHAN—Manager, Business Strategy, Ford of Europe, U.K.
CARLOS CALLEJA PINEDO—Chairman of the Board, Mac Electronica, S.A. de C.V., Mexico
MAURICIO DE MARIA Y CAMPOS—Subsecretaria de Fomento Industrial, Secretario de Comercio y Fomento Industrial, Government of Mexico
FRANÇOIS CASTAING—Vice President, Vehicle Engineering, Chrysler Corporation
JAY CHAI—Executive Vice President, C. Itoh & Co. (America) Inc.
CHEN ZUTAO—Chairman, China National Automotive Industry Corporation, Peoples Republic of China

*Titles of individuals are shown as they were at the time of participation in IMVP.

JUNE-SUK CHOO—Director, Industrial Policy Division, Ministry of Trade and Industry, Republic of Korea
MICHAEL COCHLIN—Under Secretary, Department of Trade and Industry, U.K.
ROBERT DALE—Managing Director, Automotive, Lucas Industries plc, U.K.
MICHAEL DRIGGS—Special Assistant to the President for Policy Development, The White House, Washington, D.C.
MANUEL DE LA PORTILLA—Director, Comercial Transmisiones y Equipos Mecanicos, S.A. de C.V., Mexico
JOHN EBY—Executive Director, Corporate Planning Office, Ford Motor Company
DONALD EPHLIN—Vice President, International Union, UAW
GUSTAVO ESPINOSA CARBAJAL—Director General, Fabrica de Autotransportes Mexicana, S.A. de C.V., Mexico
CESAR FLORES—Executive President, Asociacion Mexicana de Industria Automotriz, Mexico
JOSE ANTONIO FREIJO—Group Director, Finishes Division, Du Pont, S.A. de C.V., Mexico
PETER FRERK—Member of the Board of Management, Volkswagen AG
JULIO ALFREDO GENEL GARCIA—Director General of Industry, Secretaria de Comercio y Fomento Industrial, Government of Mexico
VITTORIO GHIDELLA—President and Chief Executive Officer, Fiat Auto SpA
ALEXANDER GIACCO—Vice President and Chief Executive Officer, Montedison SpA
JOHN GILCHRIST—Director Ejecutivo de Finanzas, Chrysler de Mexico, S.A.
KATHERINE GILLMAN—Deputy Project Director, Office of Technology Assessment, U.S. Congress
GORDON GOW—Deputy Minister, Ministry of Industry, Trade and Technology, Government of Ontario, Canada
JOHN GRANT—Executive Director, Corporate Strategy Staff, Ford Motor Company
DONALD GSCHWIND—Executive Vice President, Product Development, Chrysler Corporation
HENRIK GUSTAVSSON—Vice President, Technical Relations, Saab-Scania AB, Sweden
MICHAEL HAMMES—Vice President, International Operations, Chrysler Corporation
MICHAEL HAWLEY—Business Planning Associate, Ford of Europe, U.K.
HIROSHI HAYANO—President of Honda of Canada Manufacturing, Inc.
KAN HIGASHI—President, New United Motor Manufacturing, Inc.
LOUIS HUGHES—Vice President and Chief Financial Officer, General Motors Europe AG
MARTIN JOSEPHI—Presidente del Consejo Ejecutivo, Volkswagen de Mexico, S.A. de C.V., Mexico
KENICHI KATO—Director, Member of the Board, Toyota Motor Corporation
YOSHIKAZU KAWANA—Member of the Board of Directors, Group Director, Europe Sales, Nissan Motor Co., Ltd.
MARYANN KELLER—Vice President, Furman Selz Mager Dietz & Birney
JEAN-PIERRE KEMPER—Director General, Automagneto, S.A. de C.V., Mexico
ALEXANDER VON KEUDELL—Vice President, TRW, Inc., Federal Republic of Germany
JOHN KIRSCHEN—External Relations Department Director, Fiat Group Delegation with the European Community

## APPENDIX C

SHOHEI KURIHARA—Senior Managing Director, Toyota Motor Corporation
MICHEL LaSALLE—Assistant Deputy Minister, Ministry of Industry and Commerce, Government of Quebec, Canada
PATRICK LAVELLE—Deputy Minister, Ministry of Industry, Trade and Technology, Government of Ontario, Canada
RAYMOND LEVY—Chairman and Chief Executive Officer, Regie Nationale des Usines Renault
JOHN LEWIS—Managing Director-Designate, E. I. du Pont de Nemours & Co., Inc.
CARLOS MADRAZO—President, Corporacion Industrias Sanluis, S.A. de C.V., Mexico
JUAN IGNACIO MARTI—Director General of the Automotive Industry, Secretaria de Comercio y Fomento Industrial, Government of Mexico
GIAN PAOLO MASSA—Senior Vice President, Strategic Marketing, Fiat Auto SpA
KEN MATTHEWS—Assistant Secretary, Automotive and Chemicals, Department of Industry, Technology and Commerce, Government of Australia
SADAO MATSUMURA—Senior Managing Director, General Manager, International Operations, Akebono Brake Industry Co., Ltd.
JOHN MCANDREWS—Group Vice President, Automotive Products, E. I. du Pont de Nemours & Co., Inc.
EMILIO MENDOZA SAEB—General Director, Direcspicer S.A. de C.V., Mexico
HANS MERKLE—Chairman, Supervisory Board, Robert Bosch GmbH
PARVIZ MOKHTARI—Corporate Vice President and Assistant General Manager, Motorola, Inc.
HEINRICH VON MOLTKE—Director, Directorate General III Internal Market and Industrial Affairs, Commission of the European Communities
HIROSHI MORIYOSHI—President, Mazda R&D of North America, Inc.
HUMBERTO MOSCONI CASTILLO—Chief Executive Officer and Director General, Diesel Nacional, S.A., Mexico
KARL-HEINZ NARJES—Vice President, Commission of the European Communities
RICHARD NEROD—President and Managing Director, General Motors Mexico
YASUSADA NOBUMOTO—Chairman, Japan Auto Parts Industries Association
ROLANDO OLVERA—President, Industria Nacional de Autopartes, Mexico
WILLIAM PAZ CASTILLO—Automotive Director, Ministry of Development, Government of Venezuela
FRANÇOIS PERRIN-PELLETIER—Secretary General, Committee of Common Market Automobile Constructors
WOLFGANG PETER—Senior Director, Car Division, Daimler-Benz AG
GONZALO PEREYRA—Vice President, Champion Interamericana, Ltd. Bujias Champion de Mexico, S.A. de C.V.
KARL H. PITZ—% IG Metall, Federal Republic of Germany
WILLIAM RAFTERY—President, Motor and Equipment Manufacturers Association
GREGORIO RAMPA—Chairman, ANFIA, Italy
HERMAN REBHAN—General Secretary, International Metalworkers Federation, Switzerland
ERICK REICKERT—President and Managing Director, Chrysler Mexico, S.A.
ROBERT REILLY—Executive Director, Corporate Strategy Staff, Ford Motor Company
PEDRO RUIZ MENDOZA—Executive Vice President, Condumex Automotive, Grupo Condumex, Mexico

GUSTAV RYDMAN—Director, Policy and Industrial Development, Saab-Valmet AB, Finland
FERNANDO SANCHEZ UGARTE—Secretaria de Comercio y de Fomento Industrial, Government of Mexico
DOMINIQUE SAVEY—Director, Plans and Products, Peugeot S.A.
NICHOLAS SCHEELE—President and Managing Director, Ford Motor Company of Mexico
LOUIS SCHWEITZER—Executive Vice President, Finance and Planning, Regie Nationale des Usines Renault
WILLIAM SCALES—Chairman, Automotive Industry Authority of Australia
GERHARD SCHULMEYER—Senior Vice President, General Manager, Automotive and Industrial Electronics Group, Motorola, Inc.
HYUN DONG SHIN—Executive Advisor, Hyundai Motor Co.
WERNER SIEBERT—Chief Economist, Volkswagen AG
CLEMENTE SIGNORONI—Senior Vice President, Corporate Development and Controller, Fiat SpA
JOHN SMITH, JR.—President, General Motors Europe AG
JOHN STEPHENSON—Rover Group
HIDEO SUGIURA—Advisor, Honda Motor Company, Ltd.
TAKAO SUZUKI—Chief, Automotive Section, Machinery and Information Bureau, Ministry of International Trade and Industry, Japan
CARL-OLOF TERNRYD—Association of Swedish Automobile Manufacturers and Wholesalers
SHOICHIRO TOYODA—President, Japan Automobile Manufacturers Association
PETER TURNBULL—Managing Director, Lex Service plc, U.K.
DANIELE VERDIANI—Director, Commission of the European Communities
ROGER VINCENT—Managing Director, Bankers Trust Company
ROBERT WATKINS—Deputy Assistant Secretary, Automotive Affairs and Consumer Goods, U.S. Department of Commerce
DAN WERBIN—Executive Vice President, Volvo Car Corporation
ROBERT WHITE—President, CAW-TCA Canada
MARINA WHITMAN—Vice President and Group Executive, General Motors Corporation
JACK WITHROW—Executive Vice President, Product Development, Chrysler Corporation
SHIGENOBU YAMAMOTO—Chairman, Hino Motors, Ltd.
YANG LINCUN—Official of Department of Science and Technology Policy, State Science and Technology Commission, Peoples Republic of China
YANG SHIH-CHIEN—Director General, Industrial Development Bureau, Ministry of Economic Affairs, Republic of China
TAIZO YOKOYAMA—Director, Deputy Executive General Manager, Office of the President, Mitsubishi Motors Corporation
CARLOS ZAMBRANO—Executive Director, Grupo Industrial Ramirez, S.A., Mexico
ENRIQUE ZAMBRANO—General Director, Metalsa, Mexico
ZHU SUI YU—China National Automotive Industry Corporation, Peoples Republic of China

# IMVP Associates

HANS AHLINDER—Project Manager, Purchasing, Volvo Car Corporation
PIERO ALESSIO—Fiat Auto SpA
DAVID BECK—Managing Director, Lex Retail Group Ltd, U.K.
MAUREEN BEARD-FREEDMAN—Senior Policy Analyst, Department of Regional Industrial Expansion, Government of Canada
AL BOSLEY—Chief Engineer, Engineering Program Planning, Chrysler Corporation
LAURETTA BORSERO—Manager, Strategic Planning, Fiat Auto SpA
CHEN, LIZHI—Senior Engineer, China National Automotive Industry Corporation, Peoples Republic of China
HARRY COOK—Director, Automotive Research, Chrysler Corporation
MICHAEL DUBE—Senior Consultant, Automotive, Ministry of Industry, Trade and Technology, Government of Ontario, Canada
NEBOJSA DIVLJAN—Director of Strategic Planning, Zastava, Yugoslavia
ELIE FARAH—Industrial Consultant, Ministry of Industry and Commerce, Government of Quebec, Canada
MICHAEL FINKELSTEIN—Associate Administrator for Research and Development, National Highway Traffic Safety Administration, U.S.A.
ROBERT FITZHENRY—Vice Chairman, The Woodbridge Group, Canada
MONTGOMERY FRAZIER—Director, Sales and Marketing, TRW Automotive
SHELDON FRIEDMAN—Research Director, International Union, UAW
GERMAINE GIBARA—Alcan Ltd., Canada
SAM GINDIN—Assistant to the President, Canadian Auto Workers Union
SHINICHI GOTO—Group Manager, Toyota Motor Corporate Service of North America
BASIL HARGROVE—Assistant to the President, Canadian Auto Workers Union
CLAES-GORAN HELANDER—Quality Manager, Volvo Passenger Cars AB
JAN HELLING—Manager, Corporate Strategy, Saab Car Division, Saab-Scania AB
MARK HOGAN—New United Motor Manufacturing, Inc.
JOHN HOLLIS—Assistant Secretary General, Committee of Common Market Automobile Constructors
DAIROKU HOSOKI—Senior Executive Vice President, Corporate Liaison, Subaru of America, Inc.
JEAN HOUOT—Deputy Director for Long-Range Planning, Peugeot, S.A.
CANDACE HOWES—International Union, UAW
HANS-VIGGO VON HULSEN—Secretary General and Chief Foreign Law Department, Volkswagen AG
ALONSO IBANEZ Y DURAN—Executive Vice President, Industria Nacional de Autopartes, Mexico
TSUTOMU KAGAWA—Associate Director, Japan Automobile Manufacturers Association
STUART KEITZ—Director, Office of Automotive Industry Affairs, U.S. Department of Commerce
REMI KELLY—Head, Automotive Division, Ministry of Industry and Commerce, Government of Quebec

SUNGSHIN KIM—Chief Engineer, In-One Development Corporation, Republic of Korea
MINORU KIYOMASU—General Manager, Tokyo Research Dept., Toyota Motor Corp.
REIJIRO KUROMIZU—Assistant Corporate General Manager, Office of International Affairs, Mitsubishi Motors Corp.
GEORGE LACY—President, Ontario Centre for Automotive Parts Technology, Canada
BOERJE LENAS—Principal Administration Officer, Planning Department, Swedish National Board for Technical Development
MANUELA LEROY—Assistant Secretary, Japan Automobile Manufacturers Association, Paris, France
ED LEVITON—Senior International Economist, Motor Vehicle Division, U.S. Department of Commerce
LI SHOUZHONG—Director, Administrative Office, China National Automotive Industry Corporation, Peoples Republic of China
LI YIN HUAN—Vice Chairman, China National Automotive Industry Corporation, Peoples Republic of China
MARVIN MILLER—Senior Research Scientist, Department of Nuclear Engineering and Center for International Studies, MIT
MUSTAFA MOHATAREM—Director of Trade Analysis, General Motors Corporation
ALFRED MOUSTACCHI—Vice President of Planning and Control of Investments, Regie Nationale des Usines Renault
MARTIN NONHEBEL—Vehicles Division, Department of Trade and Industry, U.K.
INDRA NOOYI—Director, Corporate Strategy, Motorola, Inc.
JUDITH O'CONNELL—Business Policy Analyst, Automotive Industry Authority of Australia
CHARLES OU—Chief of Transportation Section, First Division Industrial Development Bureau, Ministry of Economic Affairs, Republic of China
ROGERS PEETERS—Head of Division, Internal Market and Industrial Affairs, Commission of the European Communities
CARLOTA PEREZ—Planning Office, Ministry of Industry, Government of Venezuela
MARY POWER—Vice President, Bankers Trust Co.
DORIAN PRINCE—Internal Market and Industrial Affairs, Commission of the European Communities
GUALBERTO RANIERI—Vice President, Corporate Communications, Fiat USA Inc.
DAVID REA—Director, Technology and Planning, E. I. du Pont de Nemours & Co., Inc.
GORDON RIGGS—Director, Strategic Studies, Corporate Strategy Staff, Ford Motor Company
STEPHEN ROGERS—Director of Planning, Magna International, Inc., Canada
YOSHIAKI SAEGUSA—Vice President, Nissan Motor Co., Ltd. Head, Washington Corporate Office
ROBERT SAMARCO—Assistant Secretary, Automotive, Electrical Equipment, and Consumer Products Branch, Department of Industry, Technology and Commerce, Australia
SHINICHI SETO—Manager, Operations Support Dept., Parts and Accessories Division, Hino Motors, Ltd.

# APPENDIX C

SHEN XIJIN—Senior Engineer, Information Division, China National Automotive Industry Corporation, Peoples Republic of China

SHI DINGHUAN—Department of Industry Technology, State Science and Technology Commission, Peoples Republic of China

MORIHARU SHIZUME—General Director, Japan Automobile Manufacturers Association, Paris, France

SLAWEK SKORUPINSKI—Director, Automotive Directorate, Department of Regional Industrial Expansion, Government of Canada

MARK SNOWDON—Booz Allen and Hamilton, Paris, France

STEPHEN SODERBERG—Partner, Wellington Management Co.

NICOLE SOLYOM-DEMESMAY—Deputy Secretary General, Committee of Common Market Automobile Constructors

RICHARD STROMBOTNE—U.S. National Highway Traffic Safety Administration

TAKEO TAKAMI—Deputy General Manager, International Planning Office, Honda Motor Company, Ltd.

SEIJI TANAKA—Director and General Manager, Mazda Motor Corporation Europe, R&D Representative Office, Federal Republic of Germany

SHINICHI TANAKA—Assistant to the Senior Vice President, Corporate Public Relations, American Honda Motor Co., Inc.

BENGT TIDHULT—Principal Program Manager, Swedish National Board for Technical Development

JAMES TRASK—Director, Competitive Analysis, Economics Staff, General Motors Corporation

FRED TUCKER—General Manager, Automotive and Industrial Electronics Group, Motorola, Inc.

YOSHINORI USUI—Assistant General Manager, Corporate Planning and Research Office, Toyota Motor Corporation

GERARDO LOPEZ VALADEZ—Director de la Industria Automotriz y Coordinacion, Secretaria de Comercio y Fomento Industrial, Government of Mexico

STEPHEN WALLMAN—Chief Engineer, Powertrain, Volvo Car Corporation

AL WARNER—Director, Motor Vehicles Division, Office of Automotive Industry Affairs, U.S. Department of Commerce

FRANK WHELAN—Chief Engineer, Engineering, Resources Planning and Control, Chrysler Corporation

JOHN WILLIAMSON—Business Planning Associate, Corporate Strategy Office, Ford Motor Co.

DAVID WORTS—General Manager, Japan Automobile Manufacturers Association (Canada)

MICHAEL WYNNE-HUGHES—Executive Director, Automotive Industry Authority of Australia

KENICHI YAMASHIRO—General Manager, Research Group, Office of Corporate Planning, Mazda Motor Corp.

TOSHIAKI YOSHINO—General Manager, Secretariat, Hino Motors, Ltd.

# APPENDIX D

## IMVP PUBLICATIONS LIST*

Brown, Jonathan, Brighton Business School, UK

*The Franchised Car Retailing Industry in the U.K.* (October 1988)

*The Future of Car Retailing*
with John J. Ferron, J. D. Power and Associates, USA (May 1989)

Cardozo, Javier, Science Policy Research Unit, University of Sussex, UK

*The Argentine Automotive Industry: Some Unavoidable Issues for a Re-entry Strategy* (May 1989)

Carlsson, Matts, Chalmers University of Technology, Sweden

*Challenges for Organization of Technical Functions*
with Christer Karlsson

*Next Practice in Managing Product Development*
with Christer Karlsson, European Institute for Advanced Studies in Management, Belgium (May 1988)

*Next Practice in Product Development: Integration of Technical Functions*
with Christer Karlsson

Clark, Joel P., Materials Systems Laboratory, MIT

---

*The listed items were prepared by IMVP researchers as part of Program research. Most are available as Working Papers. A few items have also been published in journals or as book chapters.

# APPENDIX D

    *Modeling Production Processes: Past Experience and Future Plans*
    with Frank R. Field III, Materials Systems Laboratory, MIT (September 1986)

    *Cost Modeling of Alternative Automobile Assembly Technologies: A Comparative Analysis*
    with Frank R. Field III, Youngun Lee, Joonchul Park and Deborah L. Thurston, Materials Science and Engineering Department, MIT (May 1987)

DesRosiers, Dennis, DesRosiers Automotive Research, Inc., Toronto, Canada

    *The Size, Structure and Performance of the Canadian Automotive Parts Industry: Identifying Critical Success Factors* (May 1987)

Ferro, Jose Roberto, Universidad Federal de Sao Carlos, Sao Carlos, Brazil, (Visiting Scholar, International Motor Vehicle Program, MIT)

    *Human Resources Management and Corporate Culture Transfer in the Brazilian Motor Vehicle Industry* (Viewgraphs only) (October 1988)

    *Strategic Alternatives for the Brazilian Motor Vehicle Industry in the 1990s* (May 1989)

Ferron, John J., J. D. Power and Associates (formerly with the National Automobile Dealers' Association, Virginia)

    *NADA's Look Ahead: Project 2000* (May 1988)

    *Distribution Dynamics* (Viewgraphs only) (October 1988)

    *The Future of Car Retailing*
    with Jonathan Brown, Brighton Business School, UK (May 1989)

Field, Frank R., III, Materials Systems Laboratory, MIT

    *Modeling Production Processes: Past Experience and Future Plans*
    with Joel P. Clark, Materials Systems Laboratory, MIT (September 1986)

    *Cost Modeling of Alternative Automobile Assembly Technologies: A Comparative Analysis*
    with Joel P. Clark, Youngun Lee, Joonchul Park and Deborah L. Thurston, Materials Science and Engineering Department, MIT (May 1987)

Fujimoto, Takahiro, Harvard Business School

    *The European Model of Product Development: Challenge and Opportunity*
    with Kim B. Clark, Harvard Business School (May 1988)

    *Consistent Patterns in Automotive Product Strategy, Product Development, and Manufacturing Performance: Road Map for the 1990s*
    with Antony Sheriff, International Motor Vehicle Program, MIT (May 1989)

Gadde, Lars-Erik, IMIT, Chalmers University of Technology, Sweden

    *Technical Development, Market Structure Development and Distribution Networks*

with Hakan Hakansson, Uppsala, and Lars-Gunnar Mattsson, Stockholm School of Economics, Sweden (September 1986)

*Industry Dynamics and Distribution*
with Lars Gunnar Mattsson, Stockholm School of Economics, Sweden (May 1987)

*Stability and Change in Automotive Distribution*
with Hakan Hakansson, Uppsala, Lars-Gunnar Mattsson and Mikael Oberg, Stockholm School of Economics, Sweden (May 1988)

*Dealer Perspectives on Manufacturers' Total Performance Strategies*
with Lars-Gunnar Mattsson, Stockholm School of Economics, Sweden (May 1989)

*Reorganizing Distribution for Total Performance—The Manufacturing Viewpoint*
with Lars-Gunnar Mattsson, Per Andersson, and Mikael Oberg, Stockholm School of Economics, Sweden (May 1989)

Graves, Andrew, Science Policy Research Unit, University of Sussex, UK

*Technology Challenges Facing the Motor Industry: Right and Wrong Strategies* (September 1986)

*Comparison of International Research and Development in the Automobile Industry*
with Daniel T. Jones, SPRU, University of Sussex, UK (September 1986)

*Comparative Trends in Automotive Research and Development* (May 1987) (revised September 1987)

*European Design and Engineering Capabilities: A Continuing Strength* (May 1988)

*Design Houses and the Introduction of New Technology: A Case Study* (May 1988) (revised October 1988)

*Technology Trends in the World Automobile Industry* (Viewgraphs only) (October 1988)

*Prometheus: A New Departure in Automobile R&D?* (May 1989)

Helper, Susan, Department of Operations Management, Boston University

*Changing Supplier Relationships in the U.S. Auto Industry: A Framework for Analysis and Proposal for Survey Research* (October 1988)

*Changing Supplier Relationships in the United States: Results of Survey Research* (May 1989)

Herrigel, Gary, Department of Political Science, MIT

*Collaborative Manufacturing: New Supplier Relations in the Automobile Industry and the Redefinition of the Industrial Corporation*
with Charles F. Sabel, Department of Political Science, MIT, and Horst Kern, University of Göttingen, West Germany (May 1989)

Heywood, John B., and Victor W. Wong, Sloan Automotive Laboratory, MIT

*A Study of How New Product Technologies Are Adopted by Automobile Companies in the United States, Europe and Japan* (September 1986)

Hyun, Young-suk, Hannam University, Korea (Visiting Scholar, International Motor Vehicle Program, MIT)

*A Technology Strategy for the Korean Motor Industry* (May 1989)

Ikeda, Masayoshi, Chuo University, Japan

*An International Comparison of Subcontracting Systems in the Automotive Component Manufacturing Industry* (May 1987)

*The Japanese Auto Component Manufacturers' System for the Division of Production* (May 1987)

*U-Line Auto Parts Production*
with Shoichiro Sei, Kanto Gakuin University, Japan, and Toshihiro Nishiguchi, International Motor Vehicle Program, MIT (October 1988)

*The Transfer of Flexible Production Systems to Japanese Auto Partsmaker Transplants in the U.S.*
with Shoichiro Sei, Kanto Gakuin University, Japan (May 1989)

Jones, Daniel T., European Research Director, International Motor Vehicle Program, Cardiff Business School, University of Wales, UK (formerly of SPRU, University of Sussex, UK)

*The Dynamics of the World Motor Vehicle Industry: Issues for Analysis* (September 1986)

*Comparison of International Research and Development in the Automobile Industry*
with Andrew Graves, Science Policy Research Unit, University of Sussex, UK (September 1986)

*Brownfields, Transplants and New Entrants: The Overcapacity Problem* (May 1987)

*Structural Adjustment in the Automobile Industry*
STI Review, OECD, No. 3, April 1988.

*The Competitive Position of the European Motor Industry: The Race for Added Value* (May 1988)

*Measuring Technological Advantage in the Motor Vehicle Industry* (May 1988)

*Key Findings of the MIT International Motor Vehicle Program 1987/88*

*A Summary of the Research Policy Forum, Spring 1988*
with Richard Lamming, Brighton Business School, UK

*The New Entrants: Searching for a Role in the World*
with James P. Womack, Research Director, International Motor Vehicle Program, MIT (May 1989)

*A Second Look at the European Motor Industry* (May 1989)

*The European Motor Industry and Japan in the 1990s*

*The JAMA Forum*, Special Issue on the Automotive Industry and EC Single Market (September 1989)

*Corporate Strategy and Technology in the Automobile Industry*
in Mark Dodgson, ed., *Technology Strategy and the Firm: Management and Public Policy*, London: Longman, 1989.

*The Competitive Outlook for the European Auto Industry*
*International Journal of Vehicle Design*, May 1990.

*Measuring Up to the Japanese: Lessons from the Motor Industry*
*University of Wales Business and Economics Review*, No. 5, 1990.

Karlsson, Christer, European Institute for Advanced Studies in Management, Brussels, Belgium

*Challenges for Organization of Technical Functions*
with Matts Carlsson, IMIT, Chalmers University of Technology, Sweden (May 1987)

*Next Practice in Managing Product Development*
with Matts Carlsson, IMIT, Chalmers University of Technology, Sweden (May 1988)

*Next Practice in Product Development: Integration of Technical Functions*
with Matts Carlsson, IMIT, Chalmers University of Technology, Sweden (May 1989)

Katz, Harry C., New York State School of Industrial and Labor Relations, Cornell University

*The Industrial Relations Challenges Facing the World Auto Industry* (May 1987)

*Effects of Industrial Relations on Productivity: Evidence from an American Auto Company*
with Thomas A. Kochan, Sloan School of Management, MIT (May 1988)

*Changing Work Practices and Productivity in the Auto Industry: A U.S.-Canada Comparison*
with Noah M. Meltz, University of Toronto, Canada (from *Industrial Relations Issues for the 1990's*, Proceedings of the 26th Conference of the Canadian Industrial Relations Association, June 4–6, 1989, Laval University, Quebec)

Klein, Hans, International Motor Vehicle Program, MIT

*Towards a U.S. National Program in Intelligent Vehicle/Highway Systems* (May 1989)

Kochan, Thomas A., Sloan School of Management, MIT

*Effects of Industrial Relations on Productivity: Evidence from an American Auto Company*
with Harry C. Katz, New York State School of Industrial and Labor Relations, Cornell University (May 1988)

# APPENDIX D

Krafcik, John F., International Motor Vehicle Program, MIT

*Learning from NUMMI* (September 1986)

*Comparative Manufacturing Practice: Imbalances and Implications*
with James P. Womack, Research Director, International Motor Vehicle Program, MIT (May 1987)

*Trends in International Automotive Assembly Practice* (September 1987)

*Comparative Analysis of Performance Indicators at World Auto Assembly Plants* (Master of Science thesis, Sloan School of Management, MIT, January 1988)

*European Manufacturing Practice in a World Perspective* (May 1988)

*Complexity and Flexibility in Motor Vehicle Assembly: A Worldwide Perspective* (May 1988)

*A Summary of Findings and Future Research in Manufacturing Practice* (October 1988)

*A Methodology for Assembly Plant Performance Determination* (October 1988)

*The Problem of Flexibility for the Supplier Industry: An Assembly Plant Perspective* (October 1988)

*Explaining High Performance Manufacturing: The International Automotive Assembly Plant Study*
with John Paul MacDuffie, International Motor Vehicle Program, MIT (May 1989)

*Assembly Plant Performance and Changing Market Structure in the Luxury Car Segment* (May 1989)

*A Comparative Analysis of Assembly Plant Automation* (May 1989)

*A First Look at Performance Levels at New Entrant Assembly Plants* (May 1989)

*The Team Concept: Models for Change*
with John Paul MacDuffie, International Motor Vehicle Program, MIT (*The JAMA Forum*, Vol. 7, No. 3, February 1989)

*A New Diet for U.S. Manufacturing*
Technology Review, MIT, January 28, 1989

*Triumph of the Lean Production System*
Sloan Management Review, MIT, Vol. 30, No. 1, Fall 1988

*The Effect of Design Manufacturability on Productivity and Quality: An Update on the IMVP Assembly Plant Study* (January 1990)

Kress, Donald L., Senior Advisor, International Motor Vehicle Program, MIT

*Implications of European Unification for the Motor Vehicle Industry* (May 1988)

*Key Issues in Motor Vehicle Distribution* (May 1988)

Lamming, Richard C., Brighton Business School, UK

*The International Automotive Components Industry: Customer-Supplier Relationships: Past, Present and Future* (May 1987)

*Changing Relationships in the North American Automotive Components Industry: North American/European Perspectives* (September 1987)

*Structural Options for the European Automotive Components Supplier Industry* (May 1988)

*The International Automotive Components Supply Industry: The Emerging Best Practice* (Viewgraphs only) (October 1988)

*The Post Japanese Model for International Automotive Components Supply* (October 1988)

*Tier Structures in the North American Automotive Components Industry* (October 1988)

*Key Findings of the MIT International Motor Vehicle Program 1987/88*
with Daniel T. Jones, Cardiff Business School, University of Wales, UK (October 1988)

*The International Automotive Components Supply Industry: The Emerging Best Practice* (October 1988)

*The International Automotive Components Industry: The Next "Best Practice" for Suppliers* (May 1989)

*Research and Development in the Automotive Components Suppliers of New Entrant Countries: The Prospects for Mexico* (May 1989)

*The Causes and Effects of Structural Change in the European Automotive Components Industry* (1989)

MacDuffie, John Paul, Sloan School of Management, MIT

*Industrial Relations and "Humanware": Japanese Investments in Automobile Manufacturing in the United States*
with Haruo Shimada, Department of Economics, Keio University, Japan (September 1986) (Revised for May 1987 meeting)

*The Interaction of Production Methods, Human Resources, and Technology in Manufacturing Practice* (October 1988)

*Explaining High Performance Manufacturing: The International Automotive Assembly Plant Study*
with John F. Krafcik, International Motor Vehicle Program, MIT (May 1989)

*Worldwide Trends in Production System Management: Work Systems, Factory Practice, and Human Resource Management* (May 1989)

*The Team Concept: Models for Change*
with John F. Krafcik, *The JAMA Forum*, Vol. 7, No. 3, February 1989

Marler, Dennis L., International Motor Vehicle Program, MIT

*The Post-Japanese Model of Automotive Component Supply: Selected North American Case Studies* (May 1989)

Mattsson, Lars-Gunnar, Stockholm School of Economics, Sweden

*Technical Development, Market Structure Development and Distribution Networks*
with Lars-Erik Gadde, IMIT, Chalmers University of Technology, and Hakan Hakansson, Uppsala, Sweden (September 1986)

*Industry Dynamics and Distribution*
with Lars-Erik Gadde, IMIT, Chalmers University of Technology, Sweden (May 1987)

*Stability and Change in Automotive Distribution*
with Lars-Erik Gadde, IMIT, Chalmers University of Technology, Hakan Hakansson, Uppsala, and Mikeal Oberg, Stockholm School of Economics, Sweden (May 1988)

*Dealer Perspectives on Manufacturers' Total Performance Strategies*
with Lars-Erik Gadde, IMIT, Chalmers University of Technology, Sweden (May 1989)

*Reorganizing Distribution for Total Performance—The Manufacturing Viewpoint*
with Per Andersson and Mikael Oberg, Stockholm School of Economics, and Lars-Erik Gadde, IMIT, Chalmers University of Technology, Sweden (May 1989)

Meltz, Noah M., University of Toronto, Canada

*Changing Work Practices and Productivity in the Auto Industry: A U.S.-Canada Comparison*
with Harry C. Katz, Cornell University (from *Industrial Relations Issues for the 1990's*, Proceedings of the 26th Conference of the Canadian Industrial Relations Association, June 4–6, 1989, Laval University, Canada)

Micheletti, Gian Federico, Politecnico di Torino, Italy

*Future Trends in Process Technology: Quality and Reliability Improvement with Automatic Production* (May 1988)

Miller, Roger, University of Quebec, Montreal, Canada

*New Locational Factors in the Automobile Industry* (October 1988)

*The New Locational Dynamics in the Automobile Industry: Assembly Facilities, Parts Plants and R&D Centers* (May 1989)

*Competitive Dynamics and R&D: The Locational Impacts* (1990)

Nishiguchi, Toshihiro, International Motor Vehicle Program, MIT

*Competing Systems of Automotive Components Supply: An Examination of the Japanese "Clustered Control" Model and the "Alps" Structure* (May 1987)

*New Trends in American Auto Components Supply: Is Good Management Always Culturally Bound?* (September 1987)

*Reforming Automotive Purchasing Organization in North America: Lessons for Europe?* (May 1988)

*U-Line Auto Parts Production*
with Masayoshi Ikeda, Chuo University, and Shoichiro Sei, Kanto Gakuin University, Japan (October 1988)

*Is JIT Really JIT?* (May 1989)

*Good Management Is Good Management: The Japanization of the U.S. Auto Industry*
The JAMA Forum, Vol. 7, No. 4, April 1989

O'Donnell, John P., International Motor Vehicle Program, MIT (also of the Transportation Systems Center, U.S. Dept. of Transportation)

*Competitive Product Programs and Anticipated Domestic Production and Auto-Related Employment for 1990* (September 1986)

*Brownfields, Transplants and New Entrants: The Overcapacity Problem (The U.S. Perspective)*
with Laurie Hussey, Transportation Systems Center, U.S. Department of Transportation (May 1987)

*Addendum to Brownfields, Transplants and New Entrants: The Overcapacity Problem (The U.S. Perspective)*
The North American Light Truck Market
with Laurie Hussey, Transportation Systems Center, U.S. Department of Transportation (September 1987)

*A Second Look at Developments in the U.S. Industry* (May 1989)

Oshima, Taku, Osaka City University, Japan, (Visiting Scholar, IMVP, MIT)

*Structural Comparison of the Japanese and Chinese Automobile Industries* (September 1987)

*Technology Transfer of Japanese Automakers in the United States: Mazda Motor Corporation Case Study* (May 1989)

Robertson, David, International Motor Vehicle Program, MIT

*CAD Systems in the Design Engineering Process* (May 1989)

Sabel, Charles F., Department of Political Science, MIT

*Collaborative Manufacturing: New Supplier Relations in the Automobile Industry and the Redefinition of the Industrial Corporation*
with Gary Herrigel, Department of Political Science, MIT and Horst Kern, University of Göttingen, West Germany

Sei, Shoichiro, Kanto Gakuin University, Japan

*The Electronic JIT System and Production Technology* (May 1987)

*U-Line Auto Parts Production*
with Masayoshi Ikeda, Chuo University, Japan, and Toshihiro Nishiguchi, International Motor Vehicle Program, MIT (October 1988)

# APPENDIX D

*The Transfer of Flexible Production Systems to Japanese Auto Partsmaker Transplants in the U.S.*
with Masayoshi Ikeda, Chuo University, Japan (May 1989)

Shamrakova, Luba, Consultant, International Motor Vehicle Program, MIT

*The Soviet Automotive Industry and Market in Light of the New Economic Reforms* (May 1989)

Sheriff, Antony, International Motor Vehicle Program, MIT

*The Competitive Product Position of Automobile Manufacturers: Performance and Strategies* (May 1988)

*Product Life Cycles and Their Strategic Implications to the Auto Industry* (October 1988)

*The Fragmentation of the World Motor Vehicle Market and Its Potential Impact on the Supplier Industry* (October 1988)

*Consistent Patterns in Automotive Product Strategy, Product Development, and Manufacturing Performance: Road Map for the 1990s*
with Takahiro Fujimoto, Harvard Business School (May 1989)

Shimada, Haruo, Department of Economics, Keio University, Japan

*The Economics of Humanware* (in Japanese), Tokyo: Diamond, 1989.

*Industrial Relations and "Humanware": Japanese Investments in Automobile Manufacturing in the United States*
with John Paul MacDuffie, Sloan School of Management, MIT (September 1986) (revised May 1987)

*New Economic and Human Resource Strategies: A Challenge for the Japanese Automobile Industry* (May 1989)

Shimokawa, Koichi, Hosei University, Japan

*The Study of Automotive Sales, Distribution and Service Systems and Its Further Revolution* (May 1987)

*Vitalizing Automobile Sales System Towards De-Matured Age: Changing Japanese Automobile Market and Recent Experiences of Japanese Automobile Manufacturers* (May 1988)

*Development of the Asian NICS Automobile Industry and Future Prospects of the Global Division of Labor—Japan, ROK, China-Taiwan and Thailand* (May 1989)

*The Future of Automobile Distribution: Revolution and Review of the Automobile Sales and Distribution System Under Changing Consumer Needs and Market Structure* (May 1989)

Tidd, Joseph, Science Policy Research Unit, University of Sussex, U.K.

*Next Steps in Assembly Automation* (May 1989)

Tomisawa, Konomi, Long-Term Credit Bank, Ltd., Tokyo, Japan

*Development of and Future Outlook for an International Division of Labor in the Automobile Industries of Asian NICs* (May 1987)

Wang, Kung, Dept. of Business Administration, National Central University, Taiwan

*Development Strategies for the Automobile and Parts Industry of the Republic of China* (May 1989)

Womack, James P., Research Director, International Motor Vehicle Program, MIT

*Cross-National Collaborations in the Motor Industry: Their Causes and Consequences* (September 1986)

*Moving Toward Worldwide Best Practice: Critical Choices for Countries and Companies* (May 1987)

*Comparative Manufacturing Practice: Imbalances and Implications*
with John F. Krafcik, International Motor Vehicle Program, MIT (May 1987)

*The Future Shape of the World Auto Industry: Rethinking Industry Structure* (September 1987)

*Collaboration as a Strategy for Achieving Best Practice* (September 1987)

*The Search for Best Practice* (September 1987)

*Multinational Joint Ventures in the Motor Vehicle Sector* (1987)

*Prospects for the U.S.-Mexican Relationship in the Motor Vehicle Sector* (1987)

*The Development of the Chinese Motor Vehicle Industry: Strategic Alternatives and the Role of Foreign Firms* (December 1987)

*The European Motor Industry in a World Context: Some Strategic Dilemmas* (May 1988)

*A Review of the IMVP Research Program* (Viewgraphs only) (October 1988)

*The State of the Auto Industry: Moving to the "Post-Japanese" Era* (October 1988)

*Strategies for a Post-National Motor Industry* (May 1989)

*The New Entrants: Searching for a Role in the World*
with Daniel T. Jones, European Research Director, Cardiff Business School, University of Wales, UK (May 1989)

*The Mexican Motor Industry: Strategies for the 1990's* (May 1989)

*A Post-National Auto Industry by the Year 2000*
The JAMA Forum, Vol. 8, No. 1, September 1989

Wong, Victor, Sloan Automotive Laboratory, MIT

*A Study of How New Product Technologies Are Adopted by Automobile Companies in the United States, Europe and Japan*

with John B. Heywood, Department of Mechanical Engineering, MIT (September 1986)

Xue, Qiang, International Motor Vehicle Program, MIT
*Outline of Research on the Chinese Motor Vehicle Industry* (October 1988)
*The Chinese Motor Industry: Challenges for the 1990's* (May 1989)

# Index

AC Spark Plug, 138
Aggressive selling, 67, 186
Agnelli, Giovanni, 44, 231, 234
Agnelli family, 197
Aisin Seiki, 195
Alfa Romeo, 121, 193
American Motors, 214
Arnold, Horace, 28, 32
Association of South East Asian Nations (ASEAN) countries, 271
Aston Martin, 25, 51, 65, 229
Atlanta Ford Assembly Plant, 96
Austin, Herbert, 44, 231, 233, 234
Austin Motor Company, 231
Australia, automobile industry in, 270–72
Automobile industry. *See also* European automobile industry; Japanese automobile industry; U.S. automobile industry
  in Australia, 270–72
  in Brazil, 269–70
  as cyclical, 42–43
  in developing countries, 263–70
  in East Asia, 267–69
  European innovations in, 46–47
  in Mexico, 87, 240–41, 264–67
  overcapacity crisis in, 12
  production by region, 43, 44
  union movement in, 42

Automotive assembly plants, 75–103
  classic lean production, 79–80
  classic mass production, 77–78
  comparison of mass and lean production, 75–77, 80–82
  craft production, 88–91
  in developing countries, 87
  diffusing lean production in, 82–84
  improving, 93–98
  IMVP survey of, 84–88, 91–93
  lean organization at level of, 98–100
  lean production as humanly fulfilling, 100–3
  at New United Motor Manufacturing Inc., 82–84, 101
  productivity in, 84, 85
Automotive design
  comparison of lean and mass production, 117–19
  comparison of research and development in mass versus lean production, 132–34
  consequences of lean design in marketplace, 119–26
  future in, 127–28
  lean innovation in practice, 131–32
  need for innovations in, 135–37
  product development
    around world, 110–11
    in lean production firm, 109–10

Automotive design *(cont.)*
 in mass-production firm, 104–9
 role of invention
  in lean production, 129–30
  in mass production, 128–20
 techniques of lean design, 112–17

Bentley, 132
Blanking press, 51
Blue-collar workers, 14
BMW, 65, 119, 188, 198
Bosch, 164, 166
Bosch, Robert, 62
Brazil, automobile industry in, 269–70
British Leyland, 193
 government control of, 197
 nationalization of, 234
Bromont Hyundai plant, 263
Buick, 105
Buick Century, 104
Buick Regal, 108
Buick Skylark, 106

Cadillac, 105, 106
 limousines, 209
Calvet, Jacques, 255
CAMMI (GM-Suzuki) plant, 252
Canada-U.S. Auto Pact (1965), 264
Canada-U.S. Free Trade Agreement (1989), 264
Career ladders, in lean enterprise management, 198–200
Center for Technology, Policy and Industrial Development, 4
Chandler, Alfred, 34
Changchun, 268
Channel loyalty, in lean production, 185–86
Chaplin, Charlie, 231
Chevrolet, 105, 106. *See also* General Motors
Chevrolet Celebrity, 104
Chevrolet Chevelle, 106
Chevrolet Corvair, 106
Chevrolet Vega, 129
China, automobile industry in, 268–69
Chrysler, 197, 213
 closing of Kenosha (Wis.) plant, 215
 crisis at, 238
 foreign operations of, 208, 242
 and parts supply, 156
 Pentastar program, 159
 sales of, 227
 St. Louis plant, 244
 strategic alliances of, 219
 streamlining of, 244
 supply system at, 139–40
Chrysler Europe, 121, 213
Chrysler LeBaron, 244
Citroën, 121, 234
Citroën, André, 44, 231
Clark, Kim, 63, 110, 114, 115, 155, 164
Clean sheet, 111
Cologne (Germany) Ford Plant, 38, 39, 45, 241
Communication, in lean design, 115–16
Components supply, in lean production, 146–53
Computer-aided design (CAD), 153
Craft production, 88–91
 characteristics of, 24
 drawbacks of, 25–26
 versus mass production, 29
 organization in, 24
 overtake of, by mass production, 227–28
 at Panhard et Levassor (P&L), 21–25
 production volume in, 24
 techniques of, 12–13
 tools in, 24
 in work force, 24
Customer relations, 169–91
 and channel loyalty in lean production, 185–86
 comparison of lean versus mass distribution, 186–87
 and European customer, 175–78
 future of, 187–88
 information technology and lean customer relations, 189–91
 and lean dealership, 184–85
 and lean production, 178–84
 and mass production, 170–75
Cyclicality, problem of, and transition to lean production, 247–51

Daewoo, 261, 262, 268
Dagenham (England) Ford plant, 38, 39, 45, 231, 241
Daimler, Gottlieb, 21

# INDEX

Daimler-Benz (Mercedes), 46, 188
Decentralization, in supply chain, 138–39
Deming, W. Edwards, 277
Detail-engineering, 156
Deutsche Bank, 198
Developing countries, automobile industry in, 87, 263–70
Diamond-Star (Mitsubishi-Chrysler) plant, 252
Dickenson, Gary, 108
Die-making
   lean-production approach to, 116–17
   mass-production approach to, 116
Dimensional creep, 22
Dodge Daytona, 244
Dong A, 261
Dorn, Robert, 105–6, 107, 108
Drucker, Peter, 11
du Pont, E. I., 40
du Pont, Pierre, 40
Durant, William, 39–40

East Asia, automobile production in, 267–69
Electronic-vehicle technologies, 135
Ellis, Evelyn Henry, 21–22, 23–24, 26
Ephlin, Donald, 95
European automobile industry
   distribution system in, 175–78
   and globalization, 213–15
   mass production in, 234–35, 277
   in the 1980s, 239
   production volume in, 123
   product range in, 121–22
   transition to lean production in, 253–55
European Economic Community (EEC), and Japanese investments, 255
European Free Trade Area, 267
Exclusive dealing, 170–71, 176

Fairfax (Kans.) General Motors plant, 96
Faurote, Fay, 28, 32
Ferrari, 65

Fiat, 119, 193, 197, 235, 257, 267, 269
   Mirafiori plant, 45, 46, 235
   use of tiering by, 165
Finance, in lean enterprise management, 192–98
Fisher Body Division (GM), 107
Flins (Renault) plant, 46, 235
Ford, Henry, 11, 138, 192, 211, 227–28, 276
   and concept of mass production, 26–30
   and customer relations, 170
   as isolationist, 230
   and moving assembly line, 28
   pacifism of, 35
   and part interchangeability, 27–28
Ford, Henry, II, 139, 192–93, 211
Ford Anglia, 211
Ford Capri, 271
Ford CDW 27, 212
Ford Escort, 212, 213
Ford Festiva, 212, 262
Ford "Flying Flivver," 39
Ford Model A, 26, 39, 210
Ford Model T, 26, 27, 30, 35, 37, 38, 126, 210, 227, 231
Ford Model Y, 39, 209–10
Ford Motor Company, 12
   Atlanta assembly plant, 96
   Cologne (Germany) plant, 38, 39, 45, 241
   Dagenham (England) plant, 38, 39, 45, 231, 241
   European operations of, 220
   financing of, 192–93
   Hermosillo (Mexico) plant, 87, 265–66
   Highland Park plant, 29, 30–31, 36–37, 38, 42, 62, 82, 229, 231–34
   link with Mazda, 212–13, 237, 271
   1980s crisis at, 237–38, 244
   parts design at, 58–59
   performance of, 244
   popularity of early cars, 37–38
   Q1 program, 159
   Rouge complex (Detroit), 33, 38–39, 48, 49
   sales of, 227
   supply chain in early, 139
   Trafford Park (England) plant, 228–29
   turnover at early, 42

Ford Motor Company *(cont.)*
 and use of outsourcing, 157
 wage level at early, 42
Ford Sable, 108, 212
Ford Sierra, 212
Ford Taurus, 96, 107, 108, 212
Ford Tempo, 212
Ford Topaz, 212
Ford TriMotor, 39
Ford V8, 39
Ford Valve Grinding Tool, 30
Framingham (Mass.) General Motors assembly plant, 77–78, 85–86, 244
Fuji Heavy Industries, 196
Fujimoto, Takahiro, 110–11, 114, 115, 155, 164

General Electric Plastics, 162
General Motors, 12, 37, 155, 257
 A-cars, 104, 108–9
 Corvair project, 129
 decentralization at, 41
 Fairfax (Kansas) assembly plant, 96
 five-model product range at, 41
 foreign operations of, 242
 founding of, 39
 Framingham (Mass.) assembly plant, 77–78, 85–86, 244
 G-cars, 104
 GM-10 project, 96, 104–9, 118, 129
 joint ventures of, 219, 238, 267
 in 1980s, 238–39
 outside financing for, 41
 parts design at, 58–59
 problem-solving by Sloan in early, 39–42
 and retraining of workers, 259–60
 sales of, 227
 Spear program of, 159
 streamlining of, 244
 supply chain in early, 138–39
 supply system at, 139–40
 Vega project, 129
 X-car project, 129
Geographic spread, in lean enterprise management, 200–3
GKN, 164
Global enterprise
 advantages of, 204–9
 managing, 209–13

GM-10 project, 96, 104–9, 118, 129
Goudevert, Daniel, 274
Graves, Andrew, 5
Great Britain, mass production in, 228–31
Greenhouse effect, 137

Handelsbank, 198
Harrison Radiator, 138
*Heijunka*, in lean supply, 151
Hermosillo (Mexico) plant, of Ford Motor Company, 87, 265–66
Highland Park Ford plant, 29, 30–31, 36–37, 38, 42, 62, 82, 229, 231–34
Honda
 distribution channels of, 180
 engineering design at, 129–30
 European base of, 216–17
 Marysville (Ohio) plant, 116–17, 198–99, 201, 202, 241
 North American operations of, 201, 216–17, 241
 and use of multiregional production, 205–7
 and United Auto Workers, 252–53
Honda Accord, 108, 109–10, 202, 218
Honda/Acura Legend, 216
Honda Concerto/Rover, 200
Honda NS-X, 25, 65
Hubai, 268
Hyatt Roller Bearing Company, 40
Hyundai, 261–63, 268
 Bromont (Quebec) plant, 263
Hyundai Excel, 261–62

Iacocca, Lee, 219
Indirect workers, in mass-production plants, 78
Industrial engineers, 31, 32
Information technology, and lean customer relations, 189–91
Interchangeablity, 27, 28
International Motor Vehicle Program (IMVP), 4–7
 attendees at policy forums, 7, Appendix C
 organizations contributing to, 6, Appendix A

# INDEX

research staff, 6, Appendix B
World Assembly Plant Survey, 75*n*
Invention
   in lean production, 129–30
   in mass production, 128–29
Inventory
   in mass production, 78
   in lean production, 62, 80, 95, 160–61
Invisible hand, 34
Isuzu, 196, 208, 219

Jaguar, 5, 119, 166, 204
Japanese automobile industry. *See also* Honda; Mazda; Mitsubishi; Nissan; Toyota
   automobile dealerships in, 184–85
   and brand loyalty, 185–86
   engine evolution in, 131–32
   flexible compensation in, 250
   global strategies of, 215–18
   hiring policies in, 251
   implications behind U.S. investments of, 252–53
   investment of, in U.S. auto industry, 240–42
   inward focus of, 272–75
   lifetime employment in, 54
   production volume in, 123
   product range in, 119, 121
   quality circles in, 56
   rework in, 56
   seniority system in, 54
Japanese Ministry of International Trade and Industry (MITI), 50–51
Job-control unionism, 42
Just-in-time inventory, 62, 95, 160–61

*Kaizen*, in lean supply, 149–50
Kamiyam, Shotaro, 66
Kanban system, 62, 67
Kansai Kyohokai, 153
Kanto Kyohokai, 153
Kenosha (Wis.) plant, closing of, 215
*Keiretsu* system in Japan, 194–95, 274
Kia, 261, 262, 268
Knowledge workers, 32
Koito, 61, 195

Korean auto industry, 261. *See also* Hyundai
Korean Ministry of Industry, 261, 262
Krafcik, John, 5, 13, 75*n*, 77, 82

Lamming, Richard, 5, 156
Leadership, in lean design, 112–13
Lean customer relations
   future of, 187–88
   and information technology, 189–91
Lean dealership, 184–85
Lean design
   consequences of, in marketplace, 119–26
   techniques, 112
      communications in, 115–16
      leadership, 112–13
      simultaneous development in, 116–17
      teamwork in, 113–15
Lean enterprise, 73
Lean enterprise management
   and advantages of global enterprise, 204–13
   career ladders in, 198–200
   failure of European industry in, 213–15
   finance in, 192–98
   geographic spread in, 200–3
   Japanese in, 215–18
   and multiregional enterprise, 218–22
Lean organization, at the plant level, 98–100
Lean production
   birthplace of, 49–51
   and changing consumer demand, 64–65
   channel loyalty in, 185–86
   company as community, 53–55
   components supply in, 146–48
   comparison of mass production with, 80–82
   in contrast to mass production, 13–14, 186–87
   and customer relations, 178–84
   and dealing with the customer, 66–68
   diffusing, 82–84
   encounter between mass production and, 235–36

Lean production *(cont.)*
  final assembly plant in, 55–58
  future of, 68–69
  growth of, 12
  as humanly fulfilling, 100–3
  initial misperceptions of, 236
  innovation in practice, 131–32
  invention in, 129–30
  inventory in, 80
  making transitions to, 257–75
  obstacles to, 257–75
  product development and engineering, 63–64, 109–10
  quality control in, 79
  research and development in, 132–34
  supply chain in, 58–62
  techniques of, 13, 51–53
  at Toyota Motor Company, 11, 48–69, 79–80
  transition to, in European auto industry, 253–55
  work force in, 53–55, 80
Lean supply, in practice, 148–53
Leland, Henry, 36
Levassor, Emile, 21, 22
Lex Group, 175, 176
Lifetime employment, 54
Lucas, 166

MacDuffie, John Paul, 75*n*
Manufacturing engineers, 32
Marelli, Magneti, 166
Marketplace, consequences of lean design in the, 119–26
Marysville (Ohio) Honda plant, 116–17, 198–99, 201, 202, 241
Mass production
  in America in 1955, 43
  British system of, 228–31, 233–34
  comparison of, with lean production, 80–82
  in contrast to lean production, 13–14, 186–87
  versus craft production, 29
  and customer relations, 170–75
  diffusion of, 44–47
  encounter between lean production and, 235–36
  in Europe, 234–35, 277
  Ford's concept of, 26–30
  at GM Framingham plant, 77–78, 85–86
  invention in, 128–29
  logical limits of, 38–39
  and opposition to lean production, 257–60
  organizations in, 33–35
  overtake of craft production by, 227–28
  parts design in, 140–42, 156–62
  parts supply in, 142–46
  and product development, 104–9
  product in, 37–38
  research and development in, 132–34
  rise and fall of, 21–47
  techniques of, 13
  tools in, 35–37
  work force in, 30–33, 55, 78
Mazda, 68, 186, 208, 194, 252
  distribution channels of, 180
  family management of, 196–97
  lean production at, 237
  link with Ford Motor, 212–13, 237, 271
  North American operations of, 201
Mazda Miata, 213
McKenna tariff, 228
Mercedes-Benz, 21, 65, 119, 198
Mercury Sable, 96
Mercury Tracer, 265
Mexican Auto Decree, 270
Mexico, automobile industry in, 87, 240–41, 264–67
MG, 46, 231
Michelin family, 197
Mirafiori (Fiat) plant, 45, 46, 235
Mitsubishi Motor Company, 194
  distribution channels of, 180
  North American operations of, 201
  strategic alliances of, 219
Miyoshi, Tateomi, 109
Morris, William, 44, 231–33, 234
Motorola, 162
Multiregional Motors (MRM), 218–22
*My Years with General Motors* (Sloan), 128

Nash, 171
Neocraftsmanship, 101–2
New United Motor Manufacturing Inc.

# INDEX

(NUMMI) plant (Fremont, Calif.), 5, 77, 82–84, 219, 252
   and supplier performance, 163
   success of, 238–39
Nippondenso, 61, 62, 195
Nishiguchi, Toshihiro, 5
Nissan, 51
   distribution channels of, 180
   Mexican plant of, 240–41
   North American operations of, 201, 202–3, 241, 265
   regional supplier associations of, 153
   and United Auto Workers, 252–53
Nissan Infiniti Q45, 146, 209
Nissan Sentra, 203
Nisshin Kogyo, 146
Nobeoka, Kentaro, 6, 112
NS-X sports car (Honda), 25, 65

Oldsmobile, 105, 106
Oldsmobile Ciera, 104
Oldsmobile Cutlass Supreme, 108
Oldsmobile F-85, 106
Ohno, Taiichi, 11, 49, 51, 52, 56, 62, 63, 67, 235, 277
Opel, 234, 242
Opel Kadett, 262, 267, 270
Organization
   in craft production, 24
   in mass production, 33–35
Outsourcing, 157
Oxford Motor Company, 231

Panhard et Levassor (P&L), 22, 126
   craft-production system at, 21–25
Peugeot, 121, 165
Peugeot family, 193, 197
Pickens, T. Boone, 195
Pioneer Electric, 5
Pontiac, 98, 105
Pontiac Grand Prix, 96, 108
Pontiac LeMans, 262
Pontiac 6000, 104
Pontiac Tempest, 106
Porsche, 58, 198
Porsche/Peich family, 198
Prehardened metals, 27
Product development
   and lean production, 109–10

   and mass production, 37–38, 104–9
   in U.S. automotive industry, 245
   worldwide, 110–11
Product engineers, 32
Production smoothing, in lean supply, 151
Production volume, in craft production, 24
PSA, 119, 121, 197

Quality, 55
Quality circles, 56
Quality control, in lean production, 79
Quandt family, 198

Renault, 119, 121, 171, 193, 213, 235, 257
   North American operations of, 215
   purchase/sale of American Motors by, 214
   use of tiering by, 165
Renault, Louis, 44, 231, 234
Research and development in lean production versus mass production, 132–34
Roos, Daniel, 4
Rouge Ford complex (Detroit), 33, 38–39, 48, 49
Rover, 119, 193, 216, 234

Saab, 58, 119, 121, 204, 198
   supply chain for, 139
Saginaw Steering, 138
Saint Louis Chrysler plant, 244
Schmidt, Paul, 108
Seat, 121, 193
Seiki, Aishin, 61
Seniority, 54
Sheriff, Antony, 6, 112
Shohokai, 153
Shop-floor worker, 33
*Shusa* system, 112–15, 146–48
Siemens, 162
Simultaneous development, in lean design, 116–17
Single-sourcing, 157–59
SKF, 164
Sloan, Alfred, 11, 39–42, 58, 128–29, 138, 139, 170, 227–28, 276

Smith, Adam, 34
Smith, Jack, 238
Society of Industrial Engineers, 153
Solex, 166
Statistical process control (SPC), 153, 159–60
Steyr, 213
Studebaker, 171
Subaru, 186
Sumitomo, 68, 194, 197
Supply chain, 138–67
  bidding in, 139, 140
  components supply in lean production, 146–48
  and cost adjustments, 141
  and debugging, 144
  and decentralization, 138–39
  *heijunka* in, 151
  hurdles in lean supply, 167–68
  *kaizen* in, 149–50
  lean supply in practice, 148–53
  making running changes in, 144
  parts design in mature mass production, 140–42
  parts supply in mature mass production, 142–46
  and problem of fluctuating volume, 145
  and production smoothing, 151
  and quality control, 140–41, 144
  reforming mass-production supply systems, 156–62
  supplier associations in lean supply, 153–56
  in U.S. automotive industry, 244–45
  and value engineering, 148–49
  Western Europe as halfway station, 163–67
  and zero defects, 151–52
Suzuki, 219
Système Panhard, 21

Takarakai, 153
Teamwork, in lean design, 113–15
Tokai Kyohokai, 153
Tools
  in craft production, 24
  in mass production, 35–37
Toshihiro Nishiguchi, 162
Total quality control (TQC), 153
Toyoda, Kiichiro, 48, 53, 54

Toyota, Eiji, 11, 48–49, 62, 63, 66, 231, 235, 277
Toyota Auto, 180
Toyota Camry, 180
Toyota Celica, 180
Toyota Corolla, 180–82
Toyota Gosei, 61, 195
Toyota-GM joint-venture plant (NUMMI). *See* New United Motor Manufacturing Inc. (NUMMI) plant
Toyota Lexus LS400, 209
Toyota Motor Company, 51, 66–67, 194
  aggressive selling by, 67–68
  components supply for, 60–62
  distribution channels of, 180–81
  efficiency of, 155
  European operations of, 241
  "five why's" philosophy at, 57, 152
  Georgetown (Ky.) plant, 119
  joint venture with General Motors, 238
  lean production at Takaoka plant, 79–80
  1946 strike at, 53–54
  North American operations of, 201
  pioneering of lean production at, 11
  and quality control, 59–60
  regional supplier associations of, 153
  rise of lean production at, 48–69
  *shusa* system at, 112–15
  spin-off of supplier companies by, 195
  and United Auto Workers, 252–53
Toyota Production System, work pace in, 259
Toyota Publica, 180
Toyota Takaoka plant
  accuracy at, 81
  productivity at, 81
Toyota Supra, 180
Toyota Toyopet, 180
Toyota Townall van, 180
Toyota Vista, 180
Trabant Volkswagen plant, 267
Trafford Park (England) Ford plant, 228–29

UAW. *See* United Auto Workers
Udevalla system, 101–2

Unibody car, 46
United Auto Workers (UAW), 42, 252–53
   and lean production, 83, 100–1
United Kingdom, brand loyalty in, 185
U.S. automobile industry. *See also* Chrysler; Ford Motor Company; General Motors
   companies in, 227
   diffusion of lean production at, 244–45
   employment in, 247
   Japanese investment in, 240–42, 252–53
   and problem of cyclicality, 247–51
   product development in, 245
   production volume in, 123
   product range in, 119, 121
   supply system at, 244–45
   transition to lean production, 246
U.S. car dealerships, 171–75
   bazaar tradition in, 173–74
   changes in, 171
   inventory in, 171
Utility man, 55

Valeo group, 166
Value analysis (VA), 153, 160
Value engineering (VE), 153
   in lean supply, 148–49
Vertical integration, 33, 34, 58

Visible hand, 34, 38
Volkswagen, 119, 171, 193, 235, 257, 274
   North American operations of, 213–14, 215
   Trabant plant, 267
   Wolfsburg plant, 45, 46, 235
Volkswagen Beetle, 46
Volkswagen Fox, 269, 270
Volkswagen Polo, 267
Volvo, 119, 121, 198
   distribution system of, 175
   Udevalla plant, 101–2
Volvo Concessionaires, 175, 178

Wallenberg family, 198
Wankel rotary engine, 196, 237
Wartburg, 267
Wolfburg (VW) plant, 46, 235
Work force
   in craft production, 24
   in lean production, 53–55, 80
   in mass production, 30–33, 55, 78

Yield, 55

*Zaibatsu* system in Japan, 193–94
Zero defects, as goal in lean supply, 151–52